T5-CVB-163

A Laboratory Manual of Mammalian Anatomy and Physiology

FOURTH EDITION

SIGMUND GROLLMAN
University of Maryland

Macmillan Publishing Co., Inc.
New York
Collier Macmillan Publishers
London

Copyright © 1978, Sigmund Grollman
Printed in the United States of America

All rights reserved. No part of this book may be reproduced or transmitted in any form or by any means, electronic or mechanical, including photocopying, recording, or any information storage and retrieval system, without permission in writing from the Publisher.

Earlier editions entitled *Laboratory Manual of Human Anatomy and Physiology* copyright 1955 and 1962 by Sigmund Grollman and published by Burgess Publishing Company, Minneapolis, Minnesota. Earlier editions entitled *A Laboratory Manual of Mammalian Anatomy and Physiology* © 1964 and copyright © 1969 and 1974 by Sigmund Grollman and published by Macmillan Publishing Co., Inc.

Macmillan Publishing Co., Inc.
866 Third Avenue, New York, New York 10022

Collier Macmillan Canada, Ltd.

PRINTING 56789 YEAR 56789

ISBN 0-02-348090-4

Preface

Anatomy and physiology form the basis of contemporary medicine, and knowledge of these subjects is also necessary for those other than doctors who will be involved with human health in their life's work. This manual is designed to suit the needs of the modern nursing and paramedical curriculum and undergraduate programs in physical education, behavioral and social sciences, agricultural sciences, and prelaw.

It is customary and correct to look upon the human body as a machine; in order to understand how this machine functions (physiology), we must also have a knowledge of the various working parts (anatomy) because structure and function are inseparable. Although the body must be regarded as an integral unit, a clear understanding of the organ systems forming it is best conveyed when they are discussed separately. The order of presentation sets forth clearly and vividly the close correlation between morphology and physiology. This manual gives a complete survey of anatomy and physiology with pervading emphasis on the interworking relationship of the various parts. We hope the presentation will help the student grasp the over-all hierarchy of biological organization rather than viewing the various systems as totally separate entities.

Like the text, the new edition of the laboratory manual has been updated and revised to aid the student in acquiring a practical knowledge of basic scientific facts and to reinforce and help clarify information covered in lectures. Many new anatomical drawings have been included to make dissection easier and more understandable. In addition, many of the physiological experiments have been rewritten so that they may be performed by inexperienced students with less help from teaching assistants. Each of the physiological experiments is designed to help students develop scientific curiosity and powers of observation and to stimulate their interest and intellectual abilities. The manual is designed for use in either a year course in human anatomy and physiology or a semester course. Although specifically designed to accompany the author's text, it will also work well with other introductory human anatomy and physiology texts.

S. G.

Contents

PART II – PHYSIOLOGY

Illustrations

General Laboratory Directions

Dissection

All dissections should be carefully and neatly made. The best results are obtained by examining the organs in their natural position. When it becomes necessary to study deeper structures, push the overlaying organs aside and cut as few structures as possible.

Proceed slowly, being certain that you understand the work as it progresses. Remove no structure you may later wish to study.

The quality of laboratory work can be judged by the appearance of the dissection.

Drawings

All drawings should be made as simple outlines and stippled, unless otherwise instructed, with label lines drawn with a straightedge and parallel to the bottom of the page. Where many similar cells or structures are seen, draw a few carefully on a large scale.

The secret of accurate drawing is close observation.

Experiments

When practicable, every student will perform each experiment either individually or in cooperation with other members of the class. Some experiments will be conducted by the instructor as class demonstrations. Each student must write a description of the experiment as though he were conducting it himself. The description should include

1. The purpose of the experiment
2. A description of the apparatus
3. Results
4. A discussion in which the results are correlated with other data

Scientific Terms

The language of anatomy seems complicated to the beginner. Most terms are taken from Latin or Greek and in the original language have commonplace meanings. The easiest way, therefore, to associate structures with their scientific names is to learn the meaning of the original roots. These are found in the dictionary.

PART I

Anatomy

1

Histology

Histology is the study of tissues. In the body many cells of the same type are joined together in groups for the purpose of performing a specific function. For example, lining the digestive tract are cells that perform the function of manufacturing digestive juices, which they pour into the digestive cavity. Another example of a tissue is a group of muscle cells capable of contracting and, while doing so, bringing closer together the parts to which they are attached. Cells capable of contraction constitute the muscular tissue.

A. **Epithelial tissue** (epi = upon; thelium = to cover) covers the outer surfaces of the body and all organs, as well as lining all tubes and cavities (including the coelom, or body cavity) (Fig. 1).

   Several types of epithelial tissue are recognized and are generally named according to their shape.

   1. **Squamous epithelium** (squamous means a scale) are broad flat cells found lining the inside of the mouth cavity and covering the surface of the skin. They largely make up the mesenteries, which hold internal organs in place and also form the sheet of peritoneum that lines the abdominal and thoracic cavities.

      A prepared slide of salamander (or frog) epidermis may be used in place of fresh material. Find the material first with the medium power of the microscope; note the shape and arrangement of the cells. Now turn to high power, following directions given for use of the microscope.

      Note the shape of the compartments that make the cells. Does the boundary, or **cell membrane**, of one cell fit up against the adjacent cell? What is the shape of the deeply stained **nucleus** located near the center of the cell? The **cytoplasm** is that part of the cell between the nucleus and the cell membrane. Is it as deeply stained as the nucleus? May it be said that the nucleus has a greater affinity for staining than the cytoplasm? Are chromatin granules visible in the nucleus? Can the **nuclear membrane** be seen readily?

(A)

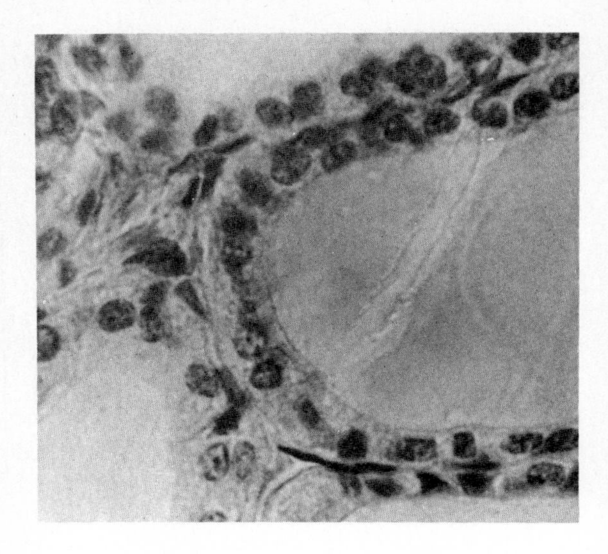

(B)

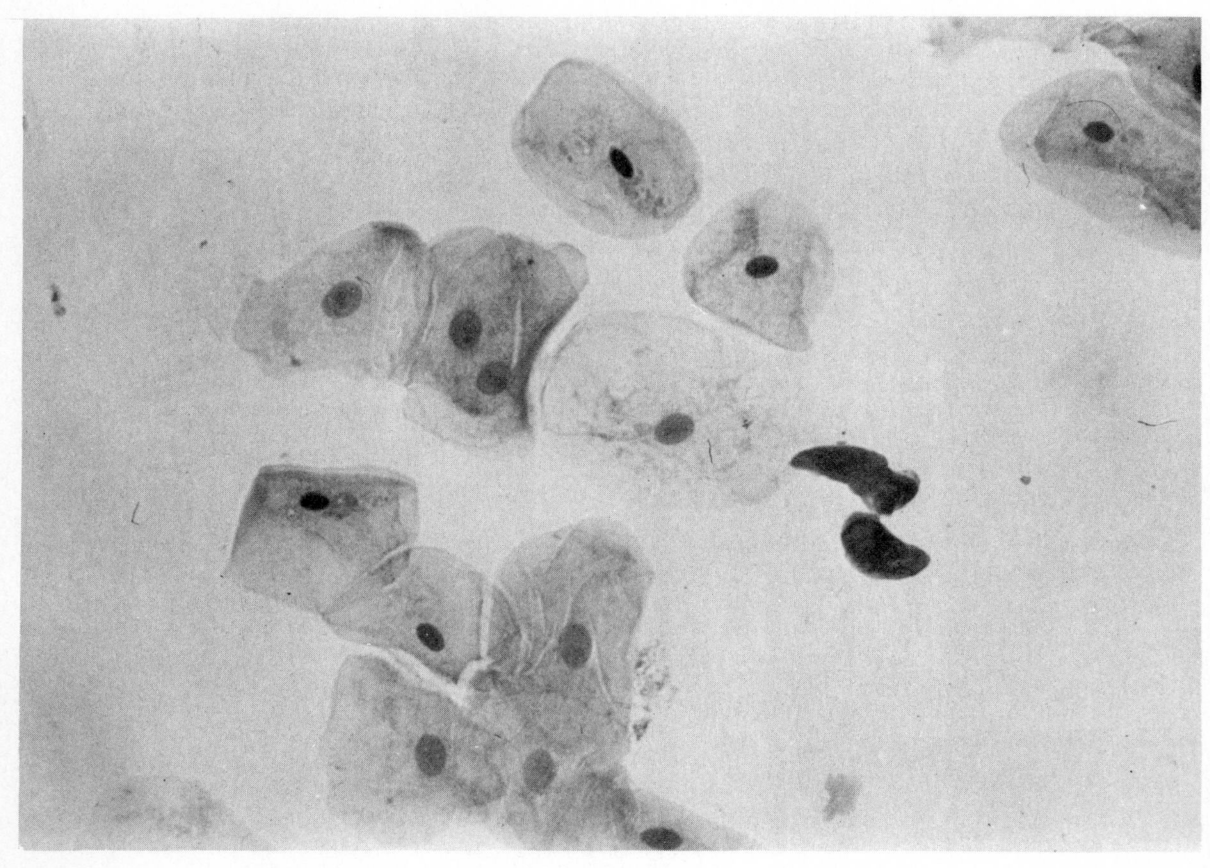

(C)

Fig. 1. Types of epithelial tissue. (Courtesy Histology Section, Zoology Department, University of Maryland, College Park, Md.)
A. Simple columnar cells in developing human tooth. 1000 X.
B. Simple cuboidal cells enclosing follicle of thyroid gland. 1000 X.
C. Squamous cells from inside mouth (cheek).

Draw a group of five or six cells three times the size they appear to you and show their relation to other cells (p. 8).

When labeling drawings, print a legend, or title, under the drawing to tell what it is, giving also the source of the material. In this case use:

SQUAMOUS EPITHELIUM SHED EPIDERMIS FROM THE FROG

Put labels of the parts of the drawing to the right of the drawing, using straight guidelines and printed labels. Here you should label the nucleus, cytoplasm, and cell membrane. Use *descriptive labels* on all drawings. This means to write a brief description of each part under the label and give its function. Use descriptive labels throughout the course.

2. **Columnar epithelium** lines the greater part of the digestive system posterior to the mouth. The cells are elongated, often tapered at their base, and broad at the surface where secretions leave the cell to pour into the digestive cavity. They are also found in other parts of the body, and in some cases, as in the lining of the trachea, are equipped with hairlike processes called **cilia**.

   Focus some of the material under low power, then turn to high power to see detail. Can you observe column shaped cells? If they are massed together, lift the cover glass and tease them apart again. The cells are wider at the top than at the bottom. Do you find some cells swollen with large vacuoles at one end? These are **goblet cells** containing secreted material about to leave the cell.

   Permanently prepared stained slides of isolated, ciliated, columnar epithelium from the trachea of the frog may be used in place of fresh material.

Draw two or three cells, each about 1½ in. long. Indicate length, width, and the amount of magnification (p. 8).

3. **Cuboidal epithelium.** In some places cuboidal (cubelike) secretory cells are grouped around a small cavity into which they pour their secretion. The alveolar glands in the skin are so constituted that the secretory portion is well protected below the surface with a duct to carry the secretion to the outside. Cells lining the kidney's uriniferous tubules are of this type. Observe a preparation of cross sectional and longitudinal sections of kidney tubules. Note the cilia on these cells. Note also the nucleus and its location and size.

Draw a section of a kidney tubule to show these features (p. 9).

B. Substantive, supportive, or connective tissue forms the framework of the body and binds organs and parts together (Fig. 2).

1. **White fibrous and yellow elastic connective tissues.** The material surrounding muscle bundles is very thin, nearly transparent, and tough. Prepare a mount of the material taken from between the muscles of the thigh of the frog. As this is picked away from

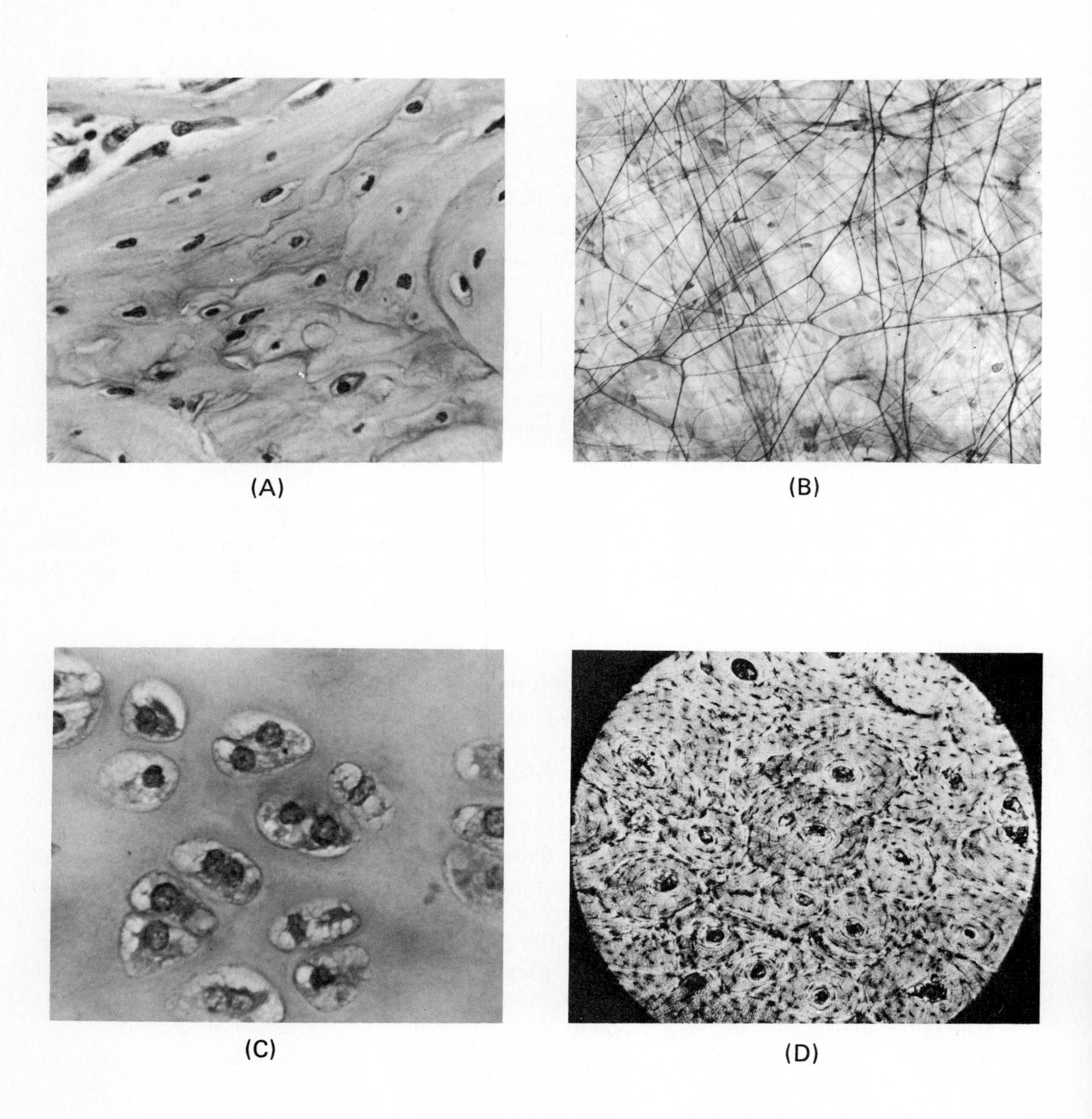

(A) (B)

(C) (D)

Fig. 2. Types of connective tissue. (Courtesy Department of Medical and Dental Illustration, Georgetown University School of Medicine, Washington, D.C.)

A. Cartilage. 800 X.
B. Areolar connective tissue showing white and yellow elastic fibers and interstitial cells.
C. Developing bone of human finger. 500 X.
D. Cross section of human bone.

the muscles with a pair of forceps, it tends to fold and curl. Place it with a drop of salt solution and spread out flat before adding the cover glass. Very thin material is better than a thick piece.

Observe under low power to obtain the focus, and then turn to high. The small fibers may be seen running throughout the material; the ones that are very fine and in wavy bundles are **white fibrous** connective tissues. The heavier, straight, and often branched fibers occurring singly are **yellow elastic** fibers. Stain for about 1 min. with methyl violet and rinse in water before remounting on a clean slide with clear water. Prepared slides of connective tissue may be used.

Draw a few fibers of each type (p. 9).

2. **Cartilage.** Cartilage is found at the ends of bones and forms the pliable framework of the ear and the tip of the nose. This material is quite transparent. The cells do not touch each other. Try to identify cells with the low power before turning to high. Do all the cells touch one another? The material between the cells is **matrix**, secreted by the cells. A **lacuna** (little lake) is the space in the matrix occupied by the cell. Are the cells found singly, in pairs, or in larger groups? Stain with methyl green to see more clearly. Prepared slides of cartilage may be used.

Draw about six cells large enough to clearly show the structure (p. 10). Give actual size and amount of magnification. Use descriptive labels on all your drawings. The matrix may be lightly shaded or left clear.

3. **Human bone.** Observe prepared slides of human bone tissue. These preparations are expensive, so be careful of them. Under low or medium power, does this appear like a group of cross sections of trees? Under high power, note in the center of each group a large cavity, or tube, generally cut crosswise. This is a **Haversian canal** (named after its discoverer). This canal contains an artery, vein, and nerve, which supply the surrounding bone area. Note model. The black dots arranged in concentric circles around the Haversian canals are the **lacunae** (little lakes), which in life were occupied by the bone cells manufacturing and secreting the hard **matrix**. The matrix is harder than that in cartilage largely because of the deposition of calcium salts in it. Look closely at one lacuna; are there tiny processes radiating in all directions from it? These tubes are generally black in microscopic preparations because they are filled with "bone dust" when the thin slices are mounted. These are **canaliculi** (little canals) (singular is canaliculus), which in life contained tiny processes from the center of cells and transported the newly manufactured matrix from the cell body to points between the cells. The Haversian canal with its surrounding lacunae constitutes a Haversian system. Bone matrix, which is hard and firm, is made of calcium and phosphorus salts; cartilage matrix is pliable.

Draw a Haversian system in outline (p. 11).

Draw one lacuna greatly enlarged showing several of the canaliculi radiating from it (p.11). Indicate size and magnification. The matrix may be lightly shaded or left clear.

Histology Drawings

SQUAMOUS EPITHELIUM

COLUMNAR EPITHELIUM

CUBOIDAL EPITHELIUM

WHITE FIBROUS AND YELLOW ELASTIC CONNECTIVE TISSUE

CARTILAGE

BONE

HAVERSIAN SYSTEM

LACUNA

2

Introduction to Anatomy

The term **anatomy** comes from the Greek, meaning "to cut up," and in its original sense it meant the knowledge of body structure gained by dissection. With the invention of the microscope around the beginning of the seventeenth century, finer details of structure were revealed, and the term "anatomy" was broadened to include microscopic structure (microscopic anatomy), which is further divided into histology (the study of tissues) and cytology (the study of cells).

Anatomy can be studied from a regional or systematic approach. In regional anatomy the body is subdivided into a number of gross regions, and the entire content of one of these (bones, muscles, nerves, blood vessels, organs) is studied before passing on to the next. In systematic anatomy the body is divided into a number of functional systems, which are studied as units. It is this approach that we use in human anatomy and physiology. The functional systems include the following:

1. The bones and their related structures, called collectively the **skeletal system**
2. The joints, or the articulatory system
3. The muscular system
4. The nervous system
5. The skin
6. The digestive system
7. The respiratory system
8. The circulatory system
9. The excretory system (urinary)
10. The reproductive system
11. The sense organs
12. The endocrine system

To orient ourselves properly with anatomic structure related to the body, it is necessary to become familiar with terms used to describe the relations of one part of the body to another. The human subject is always considered as standing upright, arms down, with palms of the hands directed forward (Fig. 3). Four common terms are

Superior: higher or upper
Inferior: under or lower

Anterior: toward the front
Posterior: toward the back

For a four legged animal the terminology is slightly different, and these terms may give rise to some confusion. The following terms avoid the difficulty by referring to the long axis of the body regardless of its position in space:

Cranial: toward the head end
Caudal: toward the tail end
Ventral: toward the animal's belly
Dorsal: toward the animal's back

These terms are also used:

Medial: toward the midline
Lateral: away from the midline
Proximal: close to the trunk
Distal: farther away from the trunk

Fig. 3. Comparison of quadruped and human showing differences in terminology relating to anatomic location.

Knee Joint (Fig. 15)

The knee is the most complex joint in the body. It not only functions in locomotion, but is the prime weight bearer of the body. Three separate articulations with a single synovial cavity cavity may be identified.

1. Articulation between the medial condyles of the **femur** (Fig. 14) and the broad surface of the **tibia**
2. Articulation between the **lateral condyles** of the femur and the **tibia**
3. Articulation between the **facets** on the posterior surface of the patella and **patellar surface** of the **femur**

The medial condyle of the femur is more distal than the lateral one, which projects forward slightly. This projection is to prevent the slipping of the patella when the quadriceps muscle contracts. The patella or kneecap is a bone within the tendon of the quadriceps femoris muscle group that seems to protect the anterior surface of the knee joint and improves the mechanical advantage of the quadriceps muscle by increasing the angle of insertion of the patellar ligament upon the tibial tuberosity. The tibia bears all the weight, as the fibula has no connection with the femur. The head of the fibula articulates with the lateral condyle of the tibia (Fig. 14), forming the **proximal tibiofibular joint.** This joint is separate from the knee joint.

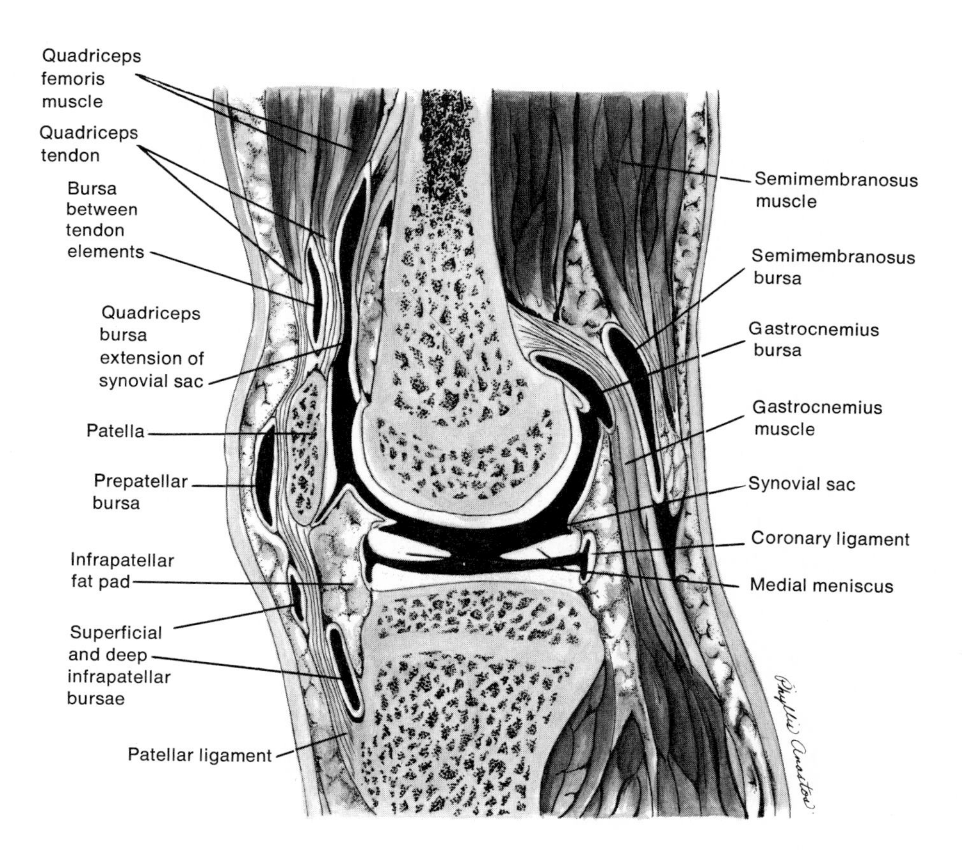

Fig. 15. Knee joint.

MUSCLES OF THE KNEE JOINT

Hamstring muscles (biceps femoris, semitendinosus, semimembranosus) – extension and adduction of thigh.
Rectus femoris – flexion of thigh at hip joint, extension of leg
Vastus lateralis – extension of knee
Vastus intermedius – extension of knee
Vastus medialis – extension of knee
Sartorius – flexion and internal rotation of leg, flexion of thigh at hip joint
Gracilis – adduction, flexion and outward rotation of thigh
Popliteus – inward rotation of leg, flexion of leg
Gastrocnemius – flexion of knee, plantar flexion and supination of foot, raising heel
Plantaris – flexion of leg at knee joint, raising heel

MOVEMENTS OF THE KNEE JOINT

Flexion, extension, inward rotation, outward rotation of the tibia

Joints of the Foot

The structure of the foot separates the human from all other members of the animal kingdom. It might be expected that the change to erect posture would be marked in the foot, which is the point of contact between the skeleton put on end and the ground, and since support and propulsion are the two functions of the human foot, the development of the longitudinal arch and transverse arches is seen.

The longitudinal arch consists of a long and short arch. The long arch is located on the inner side of the foot and extends from the calcaneus of the talus bone. The short arch is located on the outside of the foot and extends from the calcaneus to the cuboid and proximal ends of the fourth and fifth metatarsals. The transverse arch extends across the foot at the metatarsal region. The arches connect the upper bony skeleton through the tibia and fibula. Body weight is transferred from the tibia to the talus and calcaneus. There is also seen an increase of size in the big toe (hallux) and a decrease in the size of the other four toes. The big toe is held in extension and close to the other toes, which are held in flexion, both situations aiding in weight bearing and propulsion. In addition, the tarsal and metatarsal bones are found to be in closer opposition than are the wrist bones.

Ankle Joint

The ankle joint (talocrural) is formed by the articulation of the tibia and fibula with the talus. The distal end of the **tibia** and **fibula** are bound closely together by the interosseous membrane and ligaments. The **medial malleolus** of the tibia (Fig. 14) articulates with the medial malleolar facet of the talus. The **lateral malleolus** of the **fibula** articulates with the lateral malleolar facet and the **talus**. The joint formed is a hinge joint and is bound and secured by several ligaments: the **deltoid**, **talofibular** and **calcaneofibular ligaments**. These ligaments also prevent forward or backward displacement of the fibula.

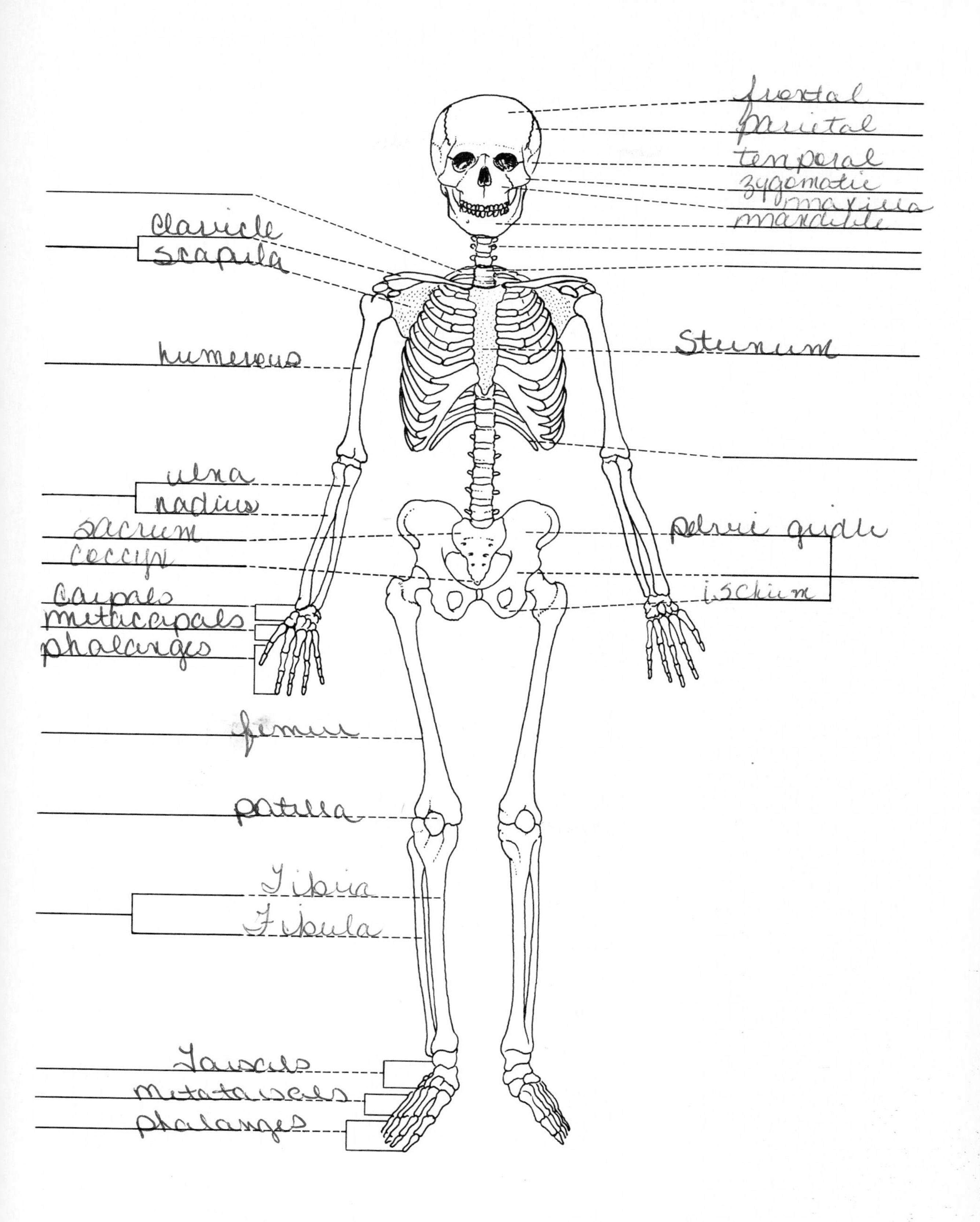

Fig. 16. Anterior view of human skelton. To be labeled by the student.

MUSCLES OF THE ANKLE JOINT

Tibialis anterior – dorsiflexion of foot at ankle joint
Extensor digitorum longus – dorsiflexion of foot at ankle joint
Peroneus tertius – eversion and plantar flexion of foot
Extensor hallucis longus – extension of great toe
Gastrocnemius – plantar flexion and supination of foot, raising heel
Plantaris – raising heel
Soleus – plantar flexion of foot, raising of heel
Peroneus longus – dorsiflexion and abduction of foot, eversion of foot
Peroneus brevis – eversion of foot
Tibialis posterior – inversion of foot
Flexor digitorum longus – flexion of phalanges of four lateral toes
Flexor hallucis longus – flexion of phalanges of "big" toe

MOVEMENTS OF THE ANKLE JOINT

Dorsiflexion – raising foot toward the anterior surface of the leg
Plantar flexion (extension) – lowering the foot so as to move it away from the anterior surface of the leg
Eversion – turning of the foot from the ankle joint laterally or outward
Inversion – turning of the foot inward or medially

OTHER JOINTS OF THE FOOT

Intertarsal joints – gliding joints between the seven tarsal bones
Tarsometatarsal joints – gliding joints formed by the articulations between the tarsal bones and the proximal ends of the five metatarsals
Metatarsophalangeal joints – joints formed between the distal ends of the metatarsals and the proximal ends of the phalanges. Flexion and extension and slight abduction and adduction are the movements of the condyloid-type joints.
Interphalangeal joints – hinge joints formed by the articulations between the phalangeal bones. Flexion and extension of the toes occur through these joints

Bones of the Foot (Fig. 14)

Tarsus (7) – ankle bones
Metatarsus (5) – bones of foot
Phalanges (14) – bones of toes

4

Histology of Muscle

Muscular, or contractile, tissue, because of its ability to contract, brings about movement of body parts and determines the ability to carry on locomotion.

Striated or Voluntary Muscle

The muscles of the thigh of the frog are composed of large, strong cells. Tear away a small mass of muscle from this part of the frog and mount in a drop of salt water. Observe with low power. Does each fiber or cell appear to have stripes (**striations**) extending crosswise? Note the model of the striated muscle. Only a short section of the cell may be visible in such a preparation. If the cells are too close, tease them apart with needles. Note the **fibrillae** running lengthwise. Turn to high power. Each fibrilla (small fiber) has thick (dark) and thin (light) bands alternately along its length. Thousands of these fibrillae in one cell have the dark and light areas opposite one another, thus giving the appearance of crosswise striations. The thin cell membrane is the **sarcolemma** (sarco = muscle; lemma = rind). Several nuclei may be present in one cell. Stain with methyl green to see the nuclei.

Draw a section of a striated muscle cell (Fig. 17)

Smooth (Unstriated) or Involuntary Muscle

The muscles of the wall of the digestive tract are made of smooth muscle fibers. A smooth muscle fiber, or cell, is long, thick in the center, and tapered at the ends. Note model. Each fibrilla is smooth like a stretched elastic. Each cell contains hundreds of lengthwise fibrillae.

Make a drawing of one smooth muscle cell (Fig. 17).

A stained permanent preparation set up as a demonstration can be used for this study.

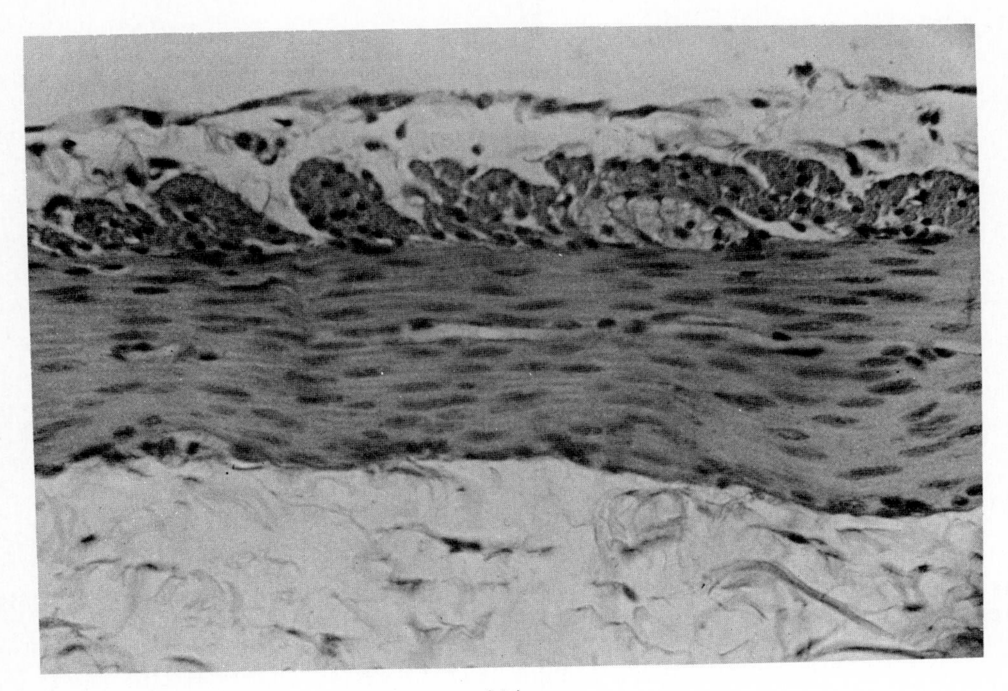

(A)

(B)

Fig. 17. Smooth and skeletal muscle. (Courtesy Department of Medical and Dental Illustration, Georgetown University, Washington, D.C.)

A. Smooth muscle – section through intestine. 158 ×

B. Skeletal muscle section showing multinucleated muscle fibers and cross striations. (Courtesy Ward's Natural Science Establishment, Inc., Rochester, N.Y.)

Histology Drawings

STRIATED MUSCLE CELLS

SMOOTH MUSCLE CELLS

5

Gross Anatomy of the Muscular System

There are many similarities between the muscular system of the cat and man. There are, however, a few differences that you will note as you dissect the cat and at the same time study the diagram of the human system in your text. Your laboratory instructor will also point out the main differences as you proceed. The points at which a skeletal muscle is attached to bones are called (1) the **origin** (stationary attachment) and (2) the **insertion** (movable attachment). By the contraction of a muscle the insertion is pulled toward the origin. A muscle may have several origins and insertions.

Muscles are attached to bones and other movable parts by connective tissue structures called tendons or aponeuroses. **Tendons** are thick, white glistening cords of closely packed collagenous fibers. It has been estimated that a large tendon would support a weight of from 10,000 to 18,000 pounds. **Aponeuroses** are sheetlike tendons. The dense connective tissue of the tendon or aponeurosis becomes continuous with the fibers of the periosteal membrane of the bone, penetrates the bone, or blends with the fibers or dermis, fibrous capsules of **joints**, or other connective tissue structures. At the muscle-tendon junction the fibers of the endomysium and perimysium fuse with the collagenous fibers of the tendon. **Ligaments** are cords of dense connective tissue that bind and give support to the various joints of the skeletal system.

Dissection of a muscle consists of breaking the fascia by means of which it is attached to neighboring structures. This is commonly done with a blunt instrument. Only the body of the muscle is separated out, and the dissection is carried as near to the origin and insertion as possible. These points of attachment are left intact so that the action of the muscle can be determined.

It is necessary to transect the superficial muscles so that the deeper ones can be studied. The muscle should be cut at right angles to the fibers and across the middle of the muscle. In this way the origin and the insertion are kept intact, and the cut edges can be brought together later to show the muscle in its original shape and position.

Study the location, origin, and insertion of the following muscles of the cat and identify them on the chart of the human muscles. Determine the action resulting from the contraction of each muscle. Refer to the description in the manual. Transect only the muscles indicated. Underlying muscles will then be exposed for study.

Dissect the muscles on the left side only. Keep the right side intact for the study of blood vessels. The skin must be removed from the body, ultimately, to study all the muscles, and it is usually convenient to remove it all at the beginning of the dissection. First, make a medial longitudinal incision through the skin on the ventral surface from the lower jaw to the tail, being careful not to cut deeper than the skin. Second, extend the incision from the midline to the arms and similarly cut into the legs. See Fig. 18 for proper guidemarks. Then, by pulling it away from the body, peel the skin and with the aid of forceps or fingers break the fascia and thin strands of subcutaneous muscle that bind the skin. You will note that as you remove the skin a thin sheath of muscle adheres to the skin. This is the **cutaneous maximus** in the body region and the **platysma** in the neck region. These generally serve to move the skin. If your specimen is a female, the mammary glands will appear as a pair of large longitudinal glandular masses along the underside of the abdomen and thorax. They should be removed and discarded. Save the skin and wrap the cat with it at the end of each laboratory period to prevent the cat from drying out.

It will be noted that a large amount of fascia and fat still cover the muscle so that the normal cleavage line between muscles cannot be seen too easily. Figure 19 shows such a cat with the skin peeled back exposing the ventral surface. Dissection of the muscles will be made easier if you remove as much fascia and fat as possible before proceeding with the dissections. The directions of the muscle fibers can be seen readily if the removal of the overlying tissue has been thoroughly done. All the fibers of one muscle are held together by a sheath of connective tissue and run in the same general direction. A neighboring muscle will have a different direction. A slight pulling action between adjacent muscles usually will expose the normal cleavage line between the muscles. Separate the muscles by picking the connective tissue between them with forceps. If the muscles separate as distinct units, the procedure is correct. If small bundles of muscle fibers appear and the muscles have a ragged appearance, you are not separating the muscle but are probably tearing a single muscle into two parts.

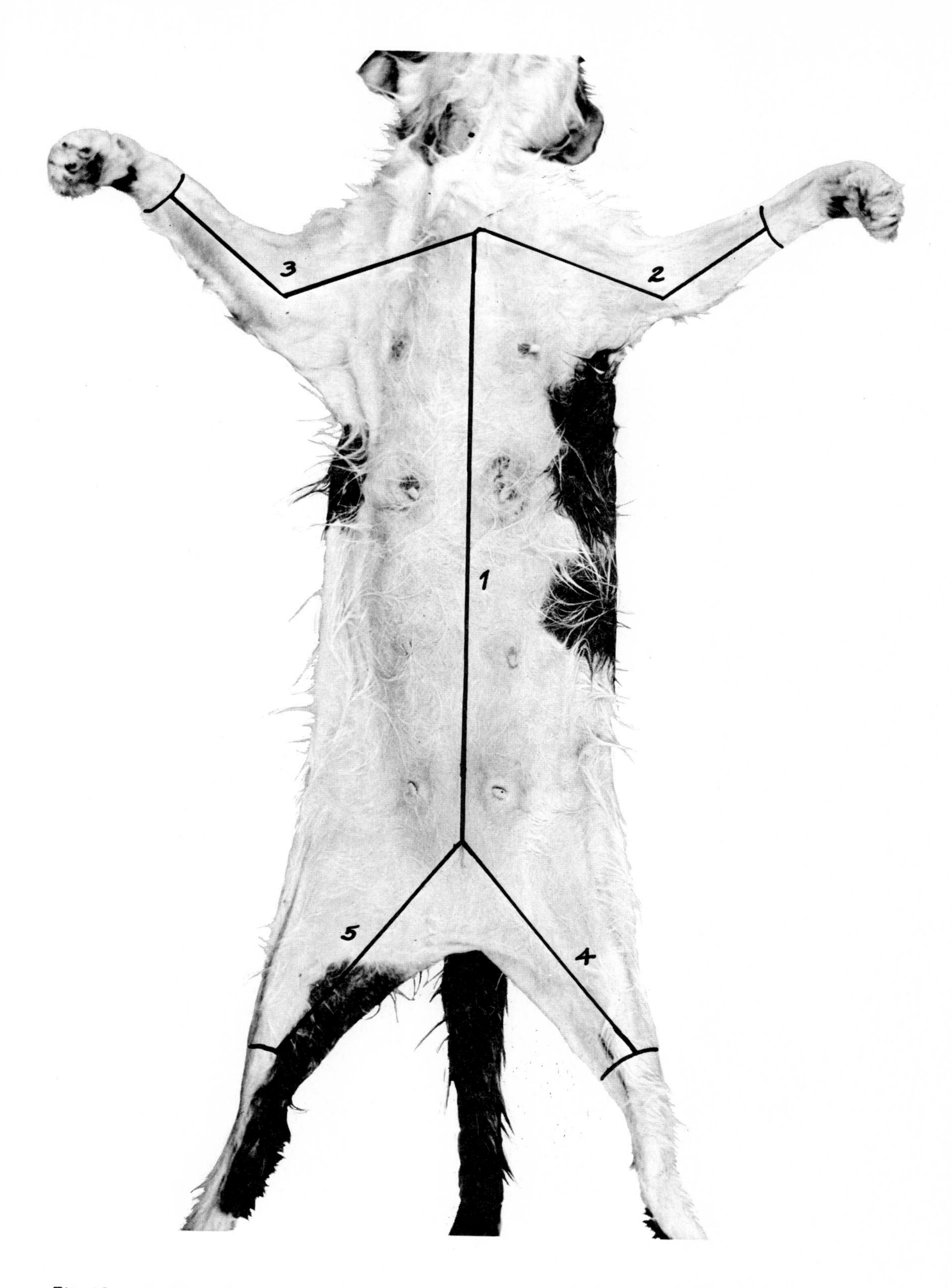

Fig. 18. Ventral surface of the cat showing the location of incisions for removing the skin.

Fig. 19. Appearance of cat when skin is reflected back from incisions. Fascia and fat are removed. Muscles become obvious and lines of separation are visible.

6

Muscles of the Chest Region, or Pectoral Group

Ventral Surface of Cat (Fig. 20)

PECTORALIS

The large triangular mass of muscle covering the chest is the **pectoralis.** It arises from the sternum and passes to insert primarily on the humerus. Its major action is to pull the arm toward the chest (adduction). In the cat the pectoralis is divided into four muscles.

1. **Pectoantibrachialis** – a band of muscle about ½ in. wide posterior to the clavodeltoid.

   Origin – front of manubrium
   Insertion – fascia at the proximal end of forearm
   Action – draws arms toward chest

2. **Pectoralis major** – posterior and deep to the pectoantibrachialis; its fibers tend to run transversely across the chest.

   Origin – middle portion of sternum
   Insertion – greater tubercular ridge and cranial border of the humerus
   Action – draws arm toward chest

3. **Pectoralis minor** – posterior to major, the greatest portion lies deep to the major. The pectoralis minor of the cat is larger than the major.

   Origin – farther on the sternum than major
   Insertion – high on the humerus
   Action – draws arm toward chest

4. **Xiphihumeralis** – most caudad of the pectoralis group. The muscle consists of a band of more or less parallel fibers that pass laterally and anteriorly and go deep to the pectoralis minor.

   Origin – xiphoid process; posterior to minor
   Insertion – proximal end of humerus; dorsal to the insertion of the other parts of the pectoralis group
   Action – draws arm toward chest

Latissimus dorsi – ventral aspect of this muscle can be seen in Fig. 20. Not a chest muscle. Will be studied later.

Clavotrapezius and clavodeltoid – also can be seen in viewing chest muscle. These are shoulder and neck muscles that will be studied later.

In man the pectoral muscle is divided into a large **pectoralis major**, a small **pectoralis minor** completely hidden by the major, and a small round muscle, the **subclavius**, which arises from the cartilage of the first rib and inserts on the undersurface of the clavicle.

Fig. 20. Muscles of the chest region, ventral surface.

1. Pectoantibrachialis.
2. Pectoralis major.
3. Pectoralis minor.
4. Xiphihumeralis.
5. Latissimus dorsi.
6. Clavotrapezius and clavodeltoid.

7

Muscles of the Abdominal Wall

Ventral View (Fig. 21)

The abdominal wall is composed of three layers of muscle plus a paired longitudinal muscle along the midventral view. All serve to compress the abdomen and thus help to elevate the diaphragm in forced expiration.

1. **External oblique** – only the dorsal half of muscle is fleshy; the ventral half is represented by flattened ribbonlike tendinous material called **aponeurosis.**

   Origin – posterior ribs and lumbodorsal fascia
   Insertion – by aponeurosis along length of linea alba
   Action – compressor of abdomen

2. **Internal oblique** – fibers of this muscle lie beneath the external oblique and extend ventrally and slightly anteriorly at right angles to the fibers of the external oblique. To expose, make a longitudinal cut through the external oblique along the lateral surface of abdomen from the costal region to the pelvic region. Do not cut deeply. With your fingers or a pair of forceps raise the edge, and watch for a deeper layer of muscle having a different direction.

   Origin – lumbodorsal fascia
   Insertion – wide aponeurosis to linea alba
   Action – compresses abdomen

3. **Rectus abdominis** – longitudinal band of muscle lying lateral to the midventral line.

   Origin – pubis
   Insertion – anterior costal cartilages and sternum
   Action – compressor of abdomen and retracts ribs and sternum

4. **Transversus abdominis** – not seen in Fig. 21. This muscle represents the third layer and lies below the internal oblique. By making a longitudinal incision through the fleshy part of the internal oblique and reflecting part of it, the transversus may be exposed. The fibers run transversely. The muscle has its origin from the cartilage of the false ribs, from the transverse processes of the lumbar vertebrae, and from the ventral border of the ilium. It inserts to the linea alba.

In man the muscle groups are similar except for the presence of an additional posterior muscle, the **quadratus lumborum**, and an anterior lateral muscle, the **pyramidalis**.

The action of abdominal muscles in man is more extensive than in the cat. They are flexors of the trunk and pelvis and side to side benders. They also help to rotate the trunk.

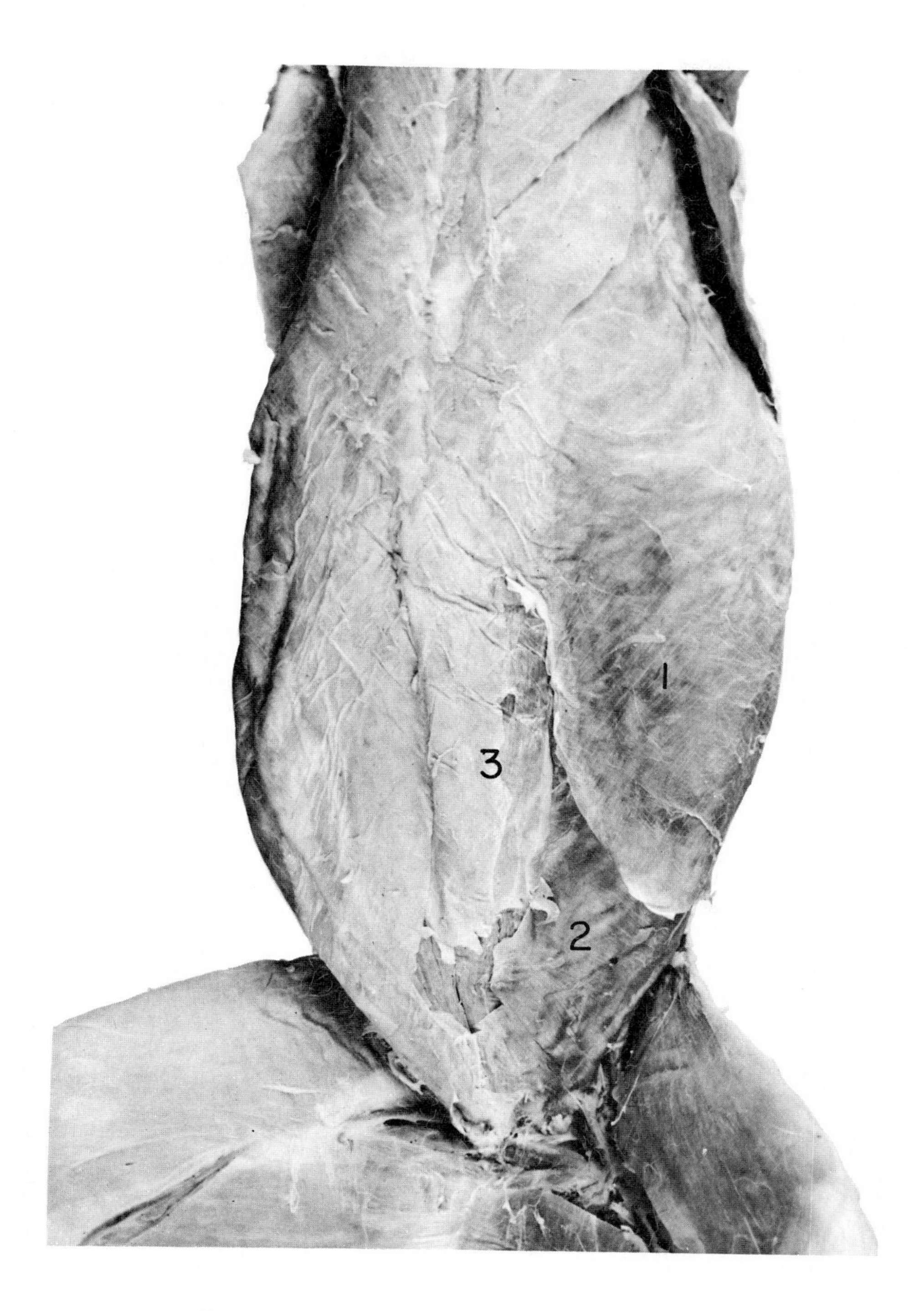

Fig. 21. Muscles of the abdominal wall, ventral surface.
1. External oblique.
2. Internal oblique.
3. Rectus abdominis.

8

Superficial Muscles of the Upper Back, Shoulders, and Back of the Neck

Dorsal Surface (Fig. 22)

TRAPEZIUS GROUPS

1. **Clavotrapezius** – clavicular portion of the trapezoid muscle in man.

   Origin – supraoccipital bone of the skull and spinous processes of first four cervical vertebrae
   Insertion – clavicle, near the proximal head of the humerus
   Action – draws clavicle dorsally and toward the head

The fibers of the clavotrapezius are continuous with those of the clavicular part of the deltoideus, which arises from the clavicle. These two muscles appear as one, which has led some to call this muscle the **cephalohumeralis**. Both work together to give a more powerful extension of the humerus.

2. **Acromiotrapezius** – large rectangular muscle that can be readily identified by its aponeurosis, which passes over the vertebral border of the scapula.

   Origin – spine of cervical and first thoracic vertebrae
   Insertion – spine of the scapula and fascia of spinotrapezius
   Action – draws scapula caudad and holds scapulae together

3. **Spinotrapezius** – triangular muscle; partly overlaps the broadest muscle of the back, the latissimus dorsi.

   Origin – spines of the thoracic vertebrae
   Insertion – fascia of scapula
   Action – draws scapula dorsally and toward the tail

In man the trapezius is one large, triangular muscle. Its name is derived from the trapezoidal figure described by the muscles of the two sides.

DELTOID GROUP – 3 MUSCLES

4. **Clavodeltoid** – appears to be part of the clavotrapezius. It is sometimes called the **clavobrachialis** because of its insertion with the brachialis.

   Origin – clavicle
   Insertion – to ulna along with the brachialis muscle
   Action – flexor of forearm

5. **Acromiodeltoid** – lies posterior to the clavodeltoid.

   Origin – acromion, deep to the levator scapulae ventralis muscle
   Insertion – deltoid tuberosity of the humerus
   Action – raises and rotates humerus along with the spinodeltoid

6. **Spinodeltoid** – lies along the lower border of the scapula posterior to the acromiotrapezius.

   Origin – spine of scapula
   Insertion – proximal end of the humerus
   Action – raises and rotates humerus

In man the deltoid is considered as one muscle that caps the point of the shoulder. It is the principal abductor of the humerus.

7. **Levator scapulae ventralis** – band of muscle that lies on the side of the neck between the sternocleidomastoid complex and the trapezius. It lies deep to the clavotrapezius.

   Origin – occipital region of skull
   Insertion – vertebral border of scapula ventral to the insertion of the acromiotrapezius
   Action – draws a scapula craniad

8. **Latissimus dorsi** – band of broad triangular muscle observed on the side of the trunk and lower part of the back posterior to the arm.

   Origin – lumbodorsal fascia and neural spines of the last thoracic vertebrae
   Insertion – proximal end of humerus
   Action – extends the humerus

In man the full action of the latissimus is seen in the crawl stroke in swimming. It is also used in all pulling movements.

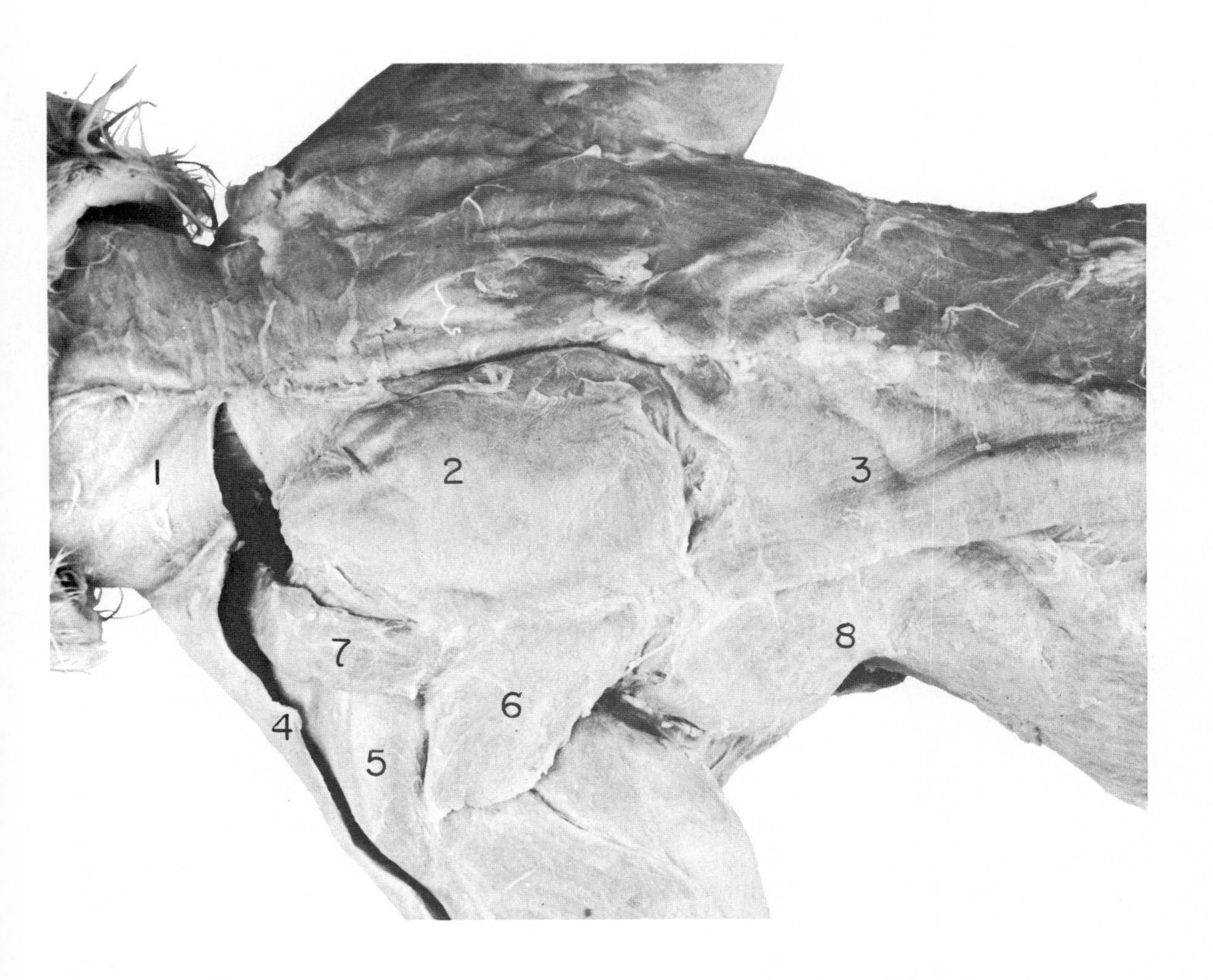

Fig. 22. Muscles of the upper back, shoulders, and back of neck, dorsal surface.

1. Clavotrapezius.
2. Acromiotrapezius.
3. Spinotrapezius.
4. Clavodeltoid.
5. Acromiodeltoid.
6. Spinodeltoid.
7. Levator scapulae ventralis.
8. Latissimus dorsi.

9

Muscles of the Upper Arm, Medial Surface

Ventral Aspect (Fig. 23)

DISSECTION

To expose the muscles of the medial surface of the upper arm, it is necessary to transect several superficial muscles of the chest group. First, transect the **pectoantibrachialis** and reflect it back to its insertion as far as possible. Next, separate the **pectoralis major** and **minor** together from the tissue lying underneath. Use your fingers to get the best separation possible. Transect both muscles and reflect. Finally, separate the latissimus as close to its origin as possible and clear away as much fascia and fat as you can. This should expose the medial surface of the upper arm sufficiently to present the muscles to be studied.

1. **Biceps brachii** – presents a prominent belly hidden by the humerus. It lies on the ventromedial surface of the humerus.

   Origin – by a tendon from the edge of the glenoid fossa. In human, second origin from coracoid, process of scapula.
   Insertion – by a tendon on the radial tuberosity of radius
   Action – powerful flexor of the forearm

2. **Epitrochlearis** – a flat band of muscle situated between the biceps and the triceps.

   Origin – by a tendon from the surface of the latissimus dorsi
   Insertion – olecranon of the ulna
   Action – extension of forearm

3. **Triceps** – this large muscle covers the posterior surface of the humerus and much of the sides. It has three main heads. The **long head** located on the posterior surface of the humerus below the epitrochlearis is the part seen in the view under study. The other parts of this muscle are seen best from the lateral surface of the upper arm and will be considered from this surface.

Long head of triceps

Origin – scapula posterior to the glenoid fossa
Insertion – olecranon process of ulna
Action – extension of forearm

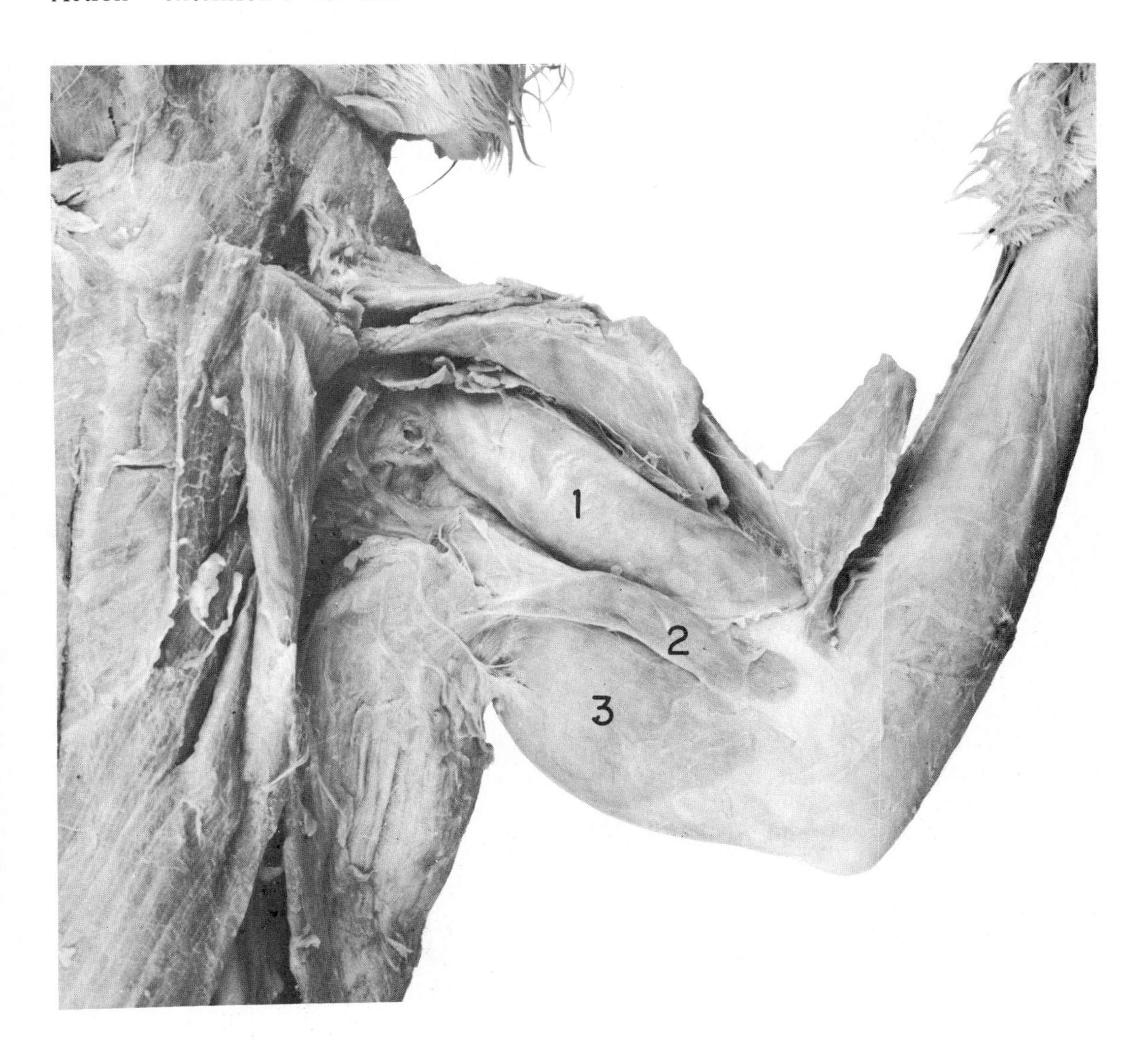

Fig. 23. Muscles of the upper arm, medial view, ventral surface.
1. Biceps brachii.
2. Epitrochlearis.
3. Triceps (long head).

10

Deep Muscles of the Chest and Medial Surface of Scapula

Ventral Surface View (Fig. 24)

Cut all the pectoral muscles and reflect back. Spread the scapula laterally away from the midventral line. This will reveal the deeper thoracic muscles and the medial surface of the scapula.

1. **Scalenus** – a complex muscle, only partly visible, consisting of three main parts: (1) dorsalis, (2) medius, (3) ventralis. Most of it lies deep in the neck muscles.

   Origin – ribs
   Insertion – transverse process of the cervical vertebrae
   Action – flexes neck or draws ribs toward head

2. **Serratus ventralis** – this muscle is composed of a series of slips, which gives it a serrated appearance.

   Origin – first nine or ten ribs (cat has 13 pairs of ribs – 9 true, 4 false of which 1 pair is floating)
   Insertion – scapula
   Action – pulls scapula forward and down

3. **Subscapularis** – large flat muscle occupying the subscapular fossa.

   Origin – surface of subscapular fossa
   Insertion – lesser tuberosity of humerus
   Action – pulls humerus medially

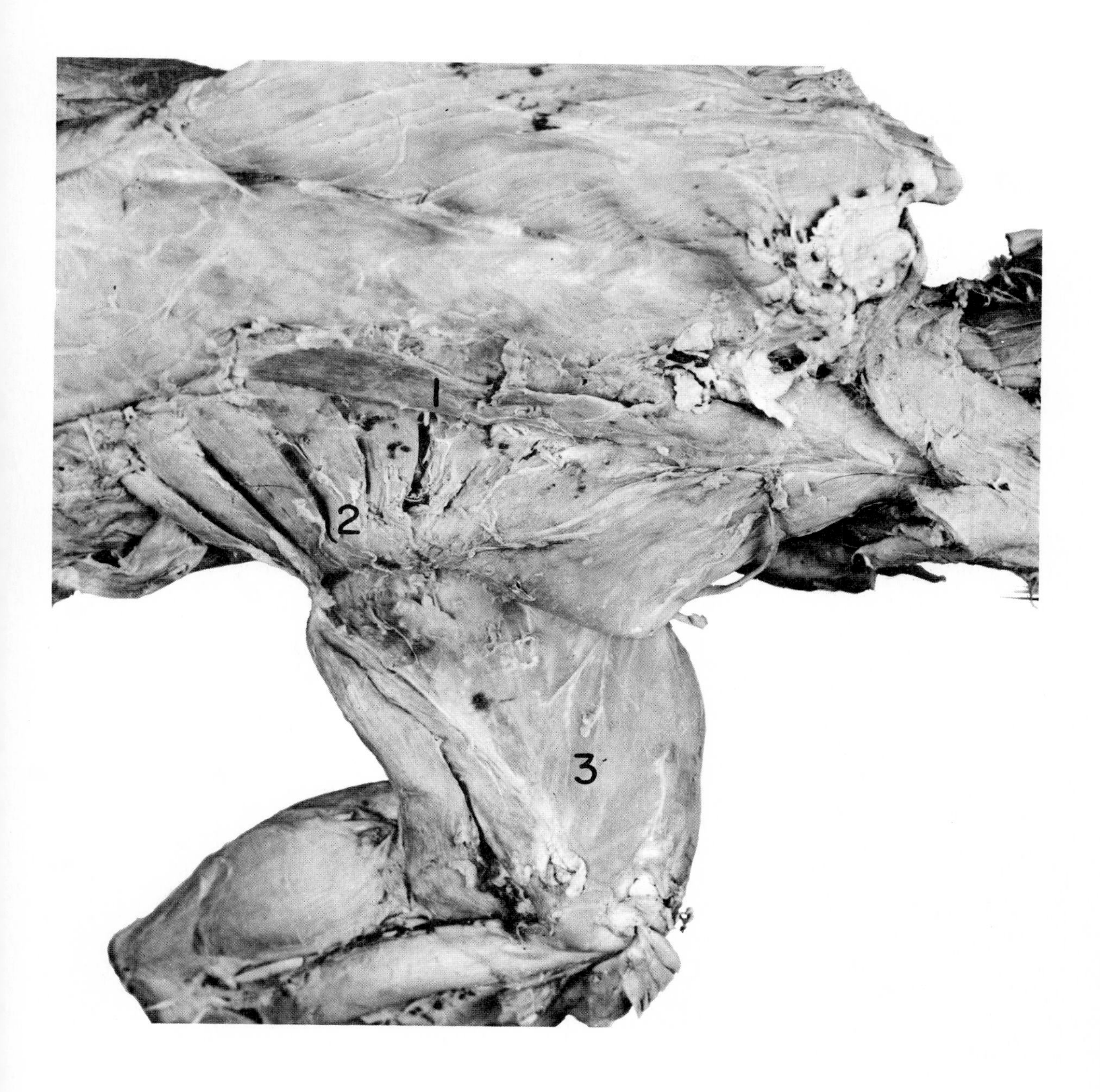

Fig. 24. Deep muscles of the chest and medial surface of the scapula, ventral view.
1. Scalenus.
2. Serratus ventralis.
3. Subscapularis.

11

Deep Muscles of the Shoulder

Dorsal Aspect (Fig. 25)

DISSECTION

Transect the latissimus and reflect the proximal portion as far dorsal as possible. Transect the spinotrapezius and remove the proximal portion by cutting it from its origin. Transect the acromiotrapezius and reflect both ends as far back as possible. Clean out the fat and loose connective tissue that presents itself as you carry out the above dissection to expose the deeper muscles of the shoulder.

1. **Supraspinatus** – large oblong muscle occupying the supraspinatus fossa. It passes over the cranial margin of the shoulder joint by a tapering tendon attached to the humerus.

   Origin – surface of supraspinatus fossa
   Insertion – greater tuberosity of humerus
   Action – extends humerus

2. **Infraspinatus** – small muscle occupying the infraspinous fossa.

   Origin – infraspinous fossa of scapula
   Insertion – greater tuberosity of humerus
   Action – rotates humerus

3. **Serratus dorsalis** – small slips of muscle lying underneath latissimus.

   Origin – aponeurosis to medial dorsal line
   Insertion – near angle to ribs
   Action – draws rib forward (anteriorly or craniad)

4. **Teres major** – band of muscle posterior to the infraspinatus and covering the axillary border of the scapula.

   Origin – axillary border of scapula
   Insertion – medial surface on proximal end of humerus
   Action – rotates humerus inward

5. **Rhomboideus** – a large muscle made up of a number of loosely associated bands.

   Origin – neural spines of posterior cervical and anterior thoracic vertebrae
   Insertion – vertebral border of scapula
   Action – draws scapula back (dorsally)

6. **Rhomboideus capitis** – not seen too well in Fig. 25 because this muscle lies deep between the scapula and body wall. It is the most anterior bundle of rhomboideus.

   Origin – back of skull
   Insertion – scapula
   Action – holds scapula in place

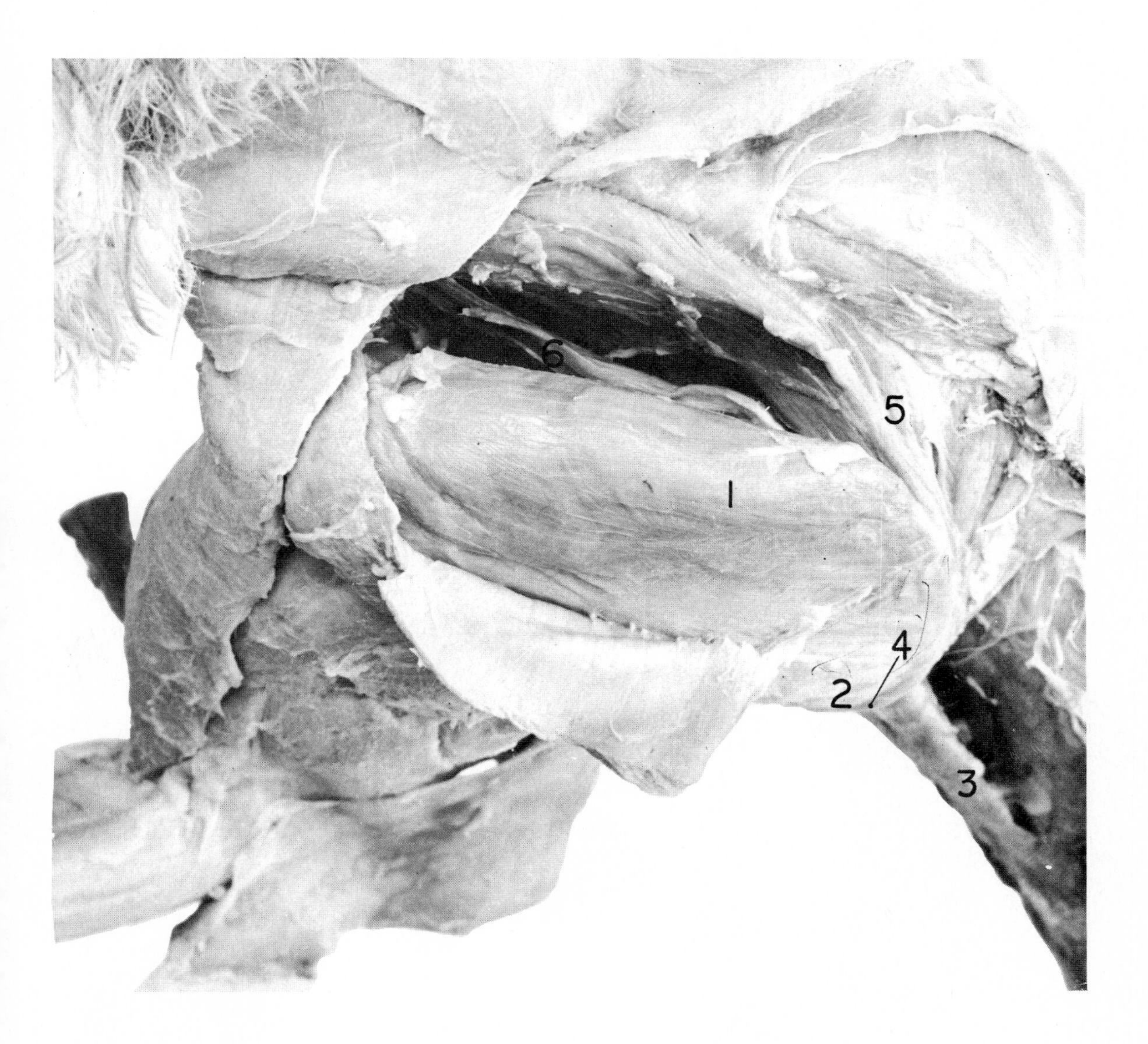

Fig. 25. Deep muscles of the shoulder, dorsal surface.

1. Supraspinatus.
2. Infraspinatus.
3. Serratus dorsalis.
4. Teres major.
5. Rhomboideus.
6. Rhomboideus capitis (not seen).

12

Deep Muscles of the Back

Dorsal Aspect (Fig. 26)

1. **Splenius** – large, thick, compactly bound muscle lying beneath the trapezius and rhomboideus capitis. There is one on either side of the neck.

   Origin – middorsal line of neck by nuchal ligament
   Insertion – lambdoidal ridge of skull
   Action – raises or turns head

2. **Semispinalis dorsi** – a small mass of muscle posterior to the splenius consisting of diagonal fibers.

   Origin – unites posteriorly and laterally with the longissimus
   Insertion – neural spines
   Action – helps maintain posture

3. **Longissimus dorsi** – elongated muscle consisting of longitudinal fibers lying posterior to the splenius and lateral to the semispinalis.

   Origin – unites in the lumbar region with the sacrospinalis
   Insertion – transverse processes of posterior cervical vertebrae
   Action – helps maintain posture

4. **Longissimus capitis** – small band of muscle surrounded by splenius, semispinalis, and longissimus dorsi.

   Origin – from the vertebrae in front of the longissimus and the spinalis
   Insertion – middorsal line
   Action – helps maintain posture

5. **Iliocostalis** – most lateral muscle of anterior back group, lying below the longissimus and deep to the fleshy portions of the serratus dorsalis.

   Origin – unites in the lumbar region with the sacrospinalis
   Insertion – ribs
   Action – extension and lateral flexion of the back, neck, and head

Lower back muscles. Reflect the lumbodorsal fascia as far back as possible. A deep layer of fascia will be seen encasing the lower back muscles. Make a longitudinal cut through it about ½ in. from the middorsal line to expose these muscles:

6. **Multifidus spinae** – narrow band of muscle beside the neural spines of the vertebrae in lumbar region.

7. **Sacrospinalis** – a wider band of muscle lateral to the multifidus. This muscle represents a united longissimus and iliocostalis of the anterior part of trunk.

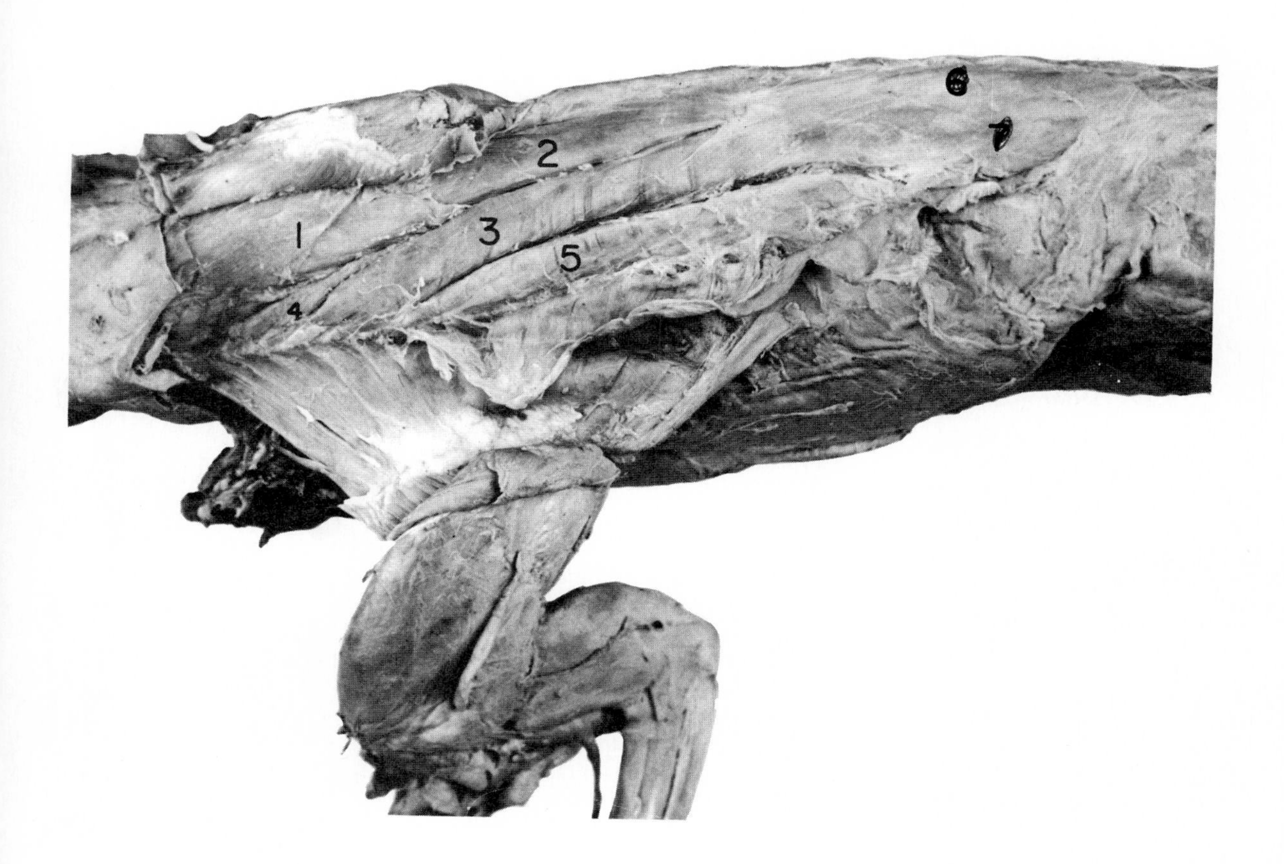

Fig. 26. Deep muscles of the back, dorsal aspect.

1. Splenius.
2. Semispinalis dorsi.
3. Longissimus dorsi.
4. Longissimus capitis.
5. Iliocostalis.
6. Multifidus spinae.
7. Sacrospinalis.

13

Muscles of the Upper Arm

Lateral Aspect (Fig. 27)

Transect the acromiodeltoideus and spinodeltoideus and reflect. This will expose the muscles of the upper arm, which act principally over the elbow joint to produce extension and flexion of the forearm.

1. **Teres minor** – a tiny round muscle caudal and adjacent to the insertion of the infraspinatus.

   Origin – distal third of the axillary border of scapula
   Insertion – greater tuberosity of humerus
   Action – rotates humerus outward

2. **Teres major** – a large round muscle posterior and ventral to the infraspinatus.

   Origin – axillary border of the scapula
   Insertion – proximal end of humerus
   Action – rotates humerus inward

Triceps brachii. A large muscle covering the posterior surface and much of the sides of the humerus. It has three main heads: a long head, a lateral head, and a medial head. In the cat a fourth head, the anconeus, is also present.

3. **Long head of triceps** – largest muscle of the group, located on the posterior surface of the humerus.

   Origin – scapula posterior to the glenoid fossa

4. **Lateral head** – large muscle covering lateral surface of humerus.

   Origin – proximal end of humerus

5. **Medial head** – small muscle mass covering medial surface of humerus and located between the long and lateral heads.

   Origin – shaft of the humerus

6. **Anconeus** – a tiny muscle lying deep to the distal end of the lateral head. It may be necessary to cut the distal end of the lateral head to see this muscle.

   Origin – distal portion of the lateral surface of the humerus
   Insertion – all parts of the triceps brachii insert in common on the olecranon process of ulna.
   Action – triceps acts to extend the forearm

7. **Brachialis** – this muscle, which covers the ventrolateral surface of the humerus, will be seen when the lateral head of the triceps is cut at its distal end and reflected.

   Origin – lateral surface of the humerus
   Insertion – ulna
   Action – flexor of forearm

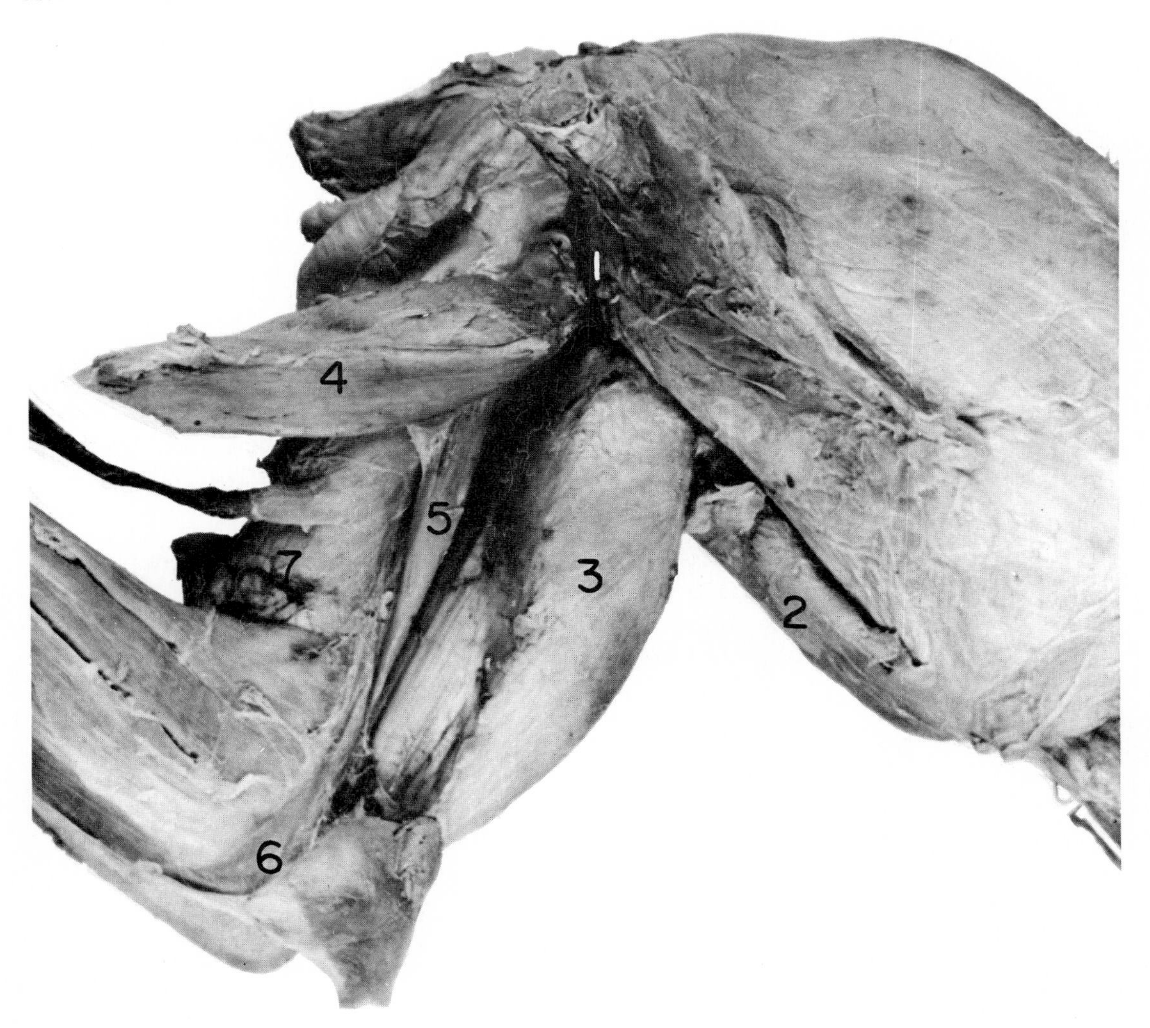

Fig. 27. Muscles of the upper arm, lateral view.
1. Teres minor.
2. Teres major.
3. Long head – triceps.
4. Lateral head – triceps.
5. Medial head – triceps.
6. Anconeus.
7. Brachialis.

14

Muscles of the Forearm

Ventral, or Palmar, Aspect (Fig. 28)

The ventral, or palmar, side of the forearm is occupied by flexors of the wrist, flexors of the digits, and special pronators of the hand. Use the brachioradialis as a landmark; then, the muscles in order from this muscle up are as follows:

1. The **extensor carpi radialis longus** (1a) and the **extensor carpi radialis brevis** (1b) represent the ventral aspect of the extensor radialis muscle.

   Origin – by common tendon attached to medial epicondyle of humerus
   Insertion – base of metacarpal (second and third)
   Action – flexion of hand at wrist joint, flexion and pronation of forearm

2. **Pronator teres** – small, triangular muscle lying above the extensor carpi radialis muscles.

   Origin – medial epicondyle of humerus
   Insertion – radius
   Action – rotates radius, pronates hand (palm down)

3. **Flexor carpi radialis** – long, spindle shaped muscle above pronator teres.

   Origin – medial epicondyle of humerus
   Insertion – base of second and third metacarpals
   Action – flexes second and third metacarpals

4. **Palmaris longus** – the broadest superficial muscle on the ventral surface. This muscle also may be used as a landmark.

   Origin – medial epicondyle of humerus
   Insertion – by tendons to phalanges
   Action – flexes phalanges

5. **Flexor carpi ulnaris** – band of muscle lying above palmaris. Consists of two heads, the first head is quite large, and the second head is very small.

   Origin – first head medial epicondyle of humerus, second head olecranon process of ulna
   Insertion – pisiform bone
   Action – flexes wrist from the ulnar side

By transecting the palmaris longus, the deep flexors of the hand can be seen. These muscles, the **flexor digitorum sublimis** and the **flexor digitorum profundis**, are not seen in the illustration.

Fig. 28. Muscles of the forearm, ventral or palmar view.
1a. Extensor carpi radialis longus.
1b. Extensor carpi ràdialis brevis.
2. Pronator teres.
3. Flexor carpi radialis.
4. Palmaris longus.
5. Flexor carpi ulnaris.

15

Muscles of the Forearm

Back, or Dorsal, View of Paw (Fig. 29)

The muscles of the forearm are involved with the movement of the hand, and in most instances the names of the muscles are descriptive of their position and particular action. Careful removal of the tough sheath of aponeurotic fascia that covers these muscles will expose them, and they can be separated from one another quite easily.

1. **Brachioradialis** – this slender band of muscle, which extends from the humerus to the lower arm and hand, can be used as a landmark to identify the other muscles.

   Origin – middle of humerus
   Insertion – lower end of radius
   Action – supinates hand

2. **Extensor carpi radialis longus** – slip of muscle next to the brachioradialis.

   Origin – humerus
   Insertion – base of second and third metacarpals
   Action – extends hand

3. **Extensor digitorum communis** – next in line to the radialis.

   Origin – lateral surface of humerus
   Insertion – four tendons into bases of second to fifth phalanges
   Action – extends phalanges

4. **Extensor digitorum lateralis**

   Origin – lateral epicondylar surface of humerus
   Insertion – by tendons to three or four phalanges
   Action – extends phalanges

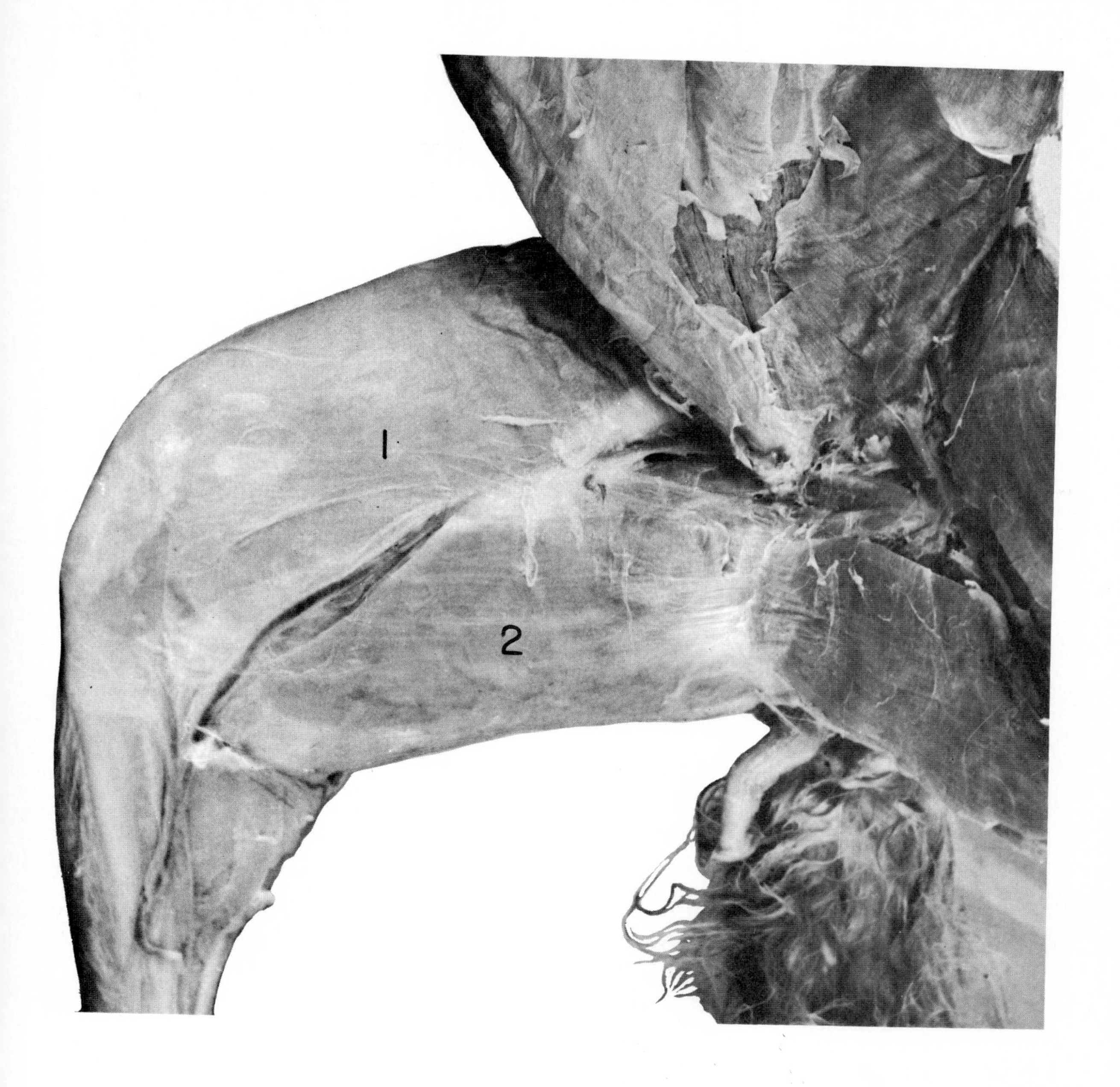

Fig. 30. Superficial muscles of the thigh, medial aspect, ventral view.
1. Sartorius.
2. Gracilis.

17

Superficial Muscles of the Thigh

Dorsal Aspect (Fig. 31)

Clean off the fat and superficial fascia from the surfaces of the pelvic region and thigh before carrying out dissection. The muscles of this group are easily separated.

1. **Gluteus medius** – a relatively large muscle in the cat and somewhat triangular.

   Origin – crest and lateral surface of ilium
   Insertion – greater trochanter of femur
   Action – aids in abducting the thigh

2. **Gluteus maximus** – triangular mass immediately posterior to the medius. It is smaller than the medius.

   Origin – last sacral and first caudal vertebrae
   Insertion – fascia lata and greater trochanter of femur
   Action – abducts thigh

3. **Caudofemoralis** – band of muscle posterior to gluteus maximus and anterior and dorsal to biceps femoris. This muscle is fused with and considered the gluteus maximus in man.

   Origin – second and third caudal vertebrae
   Insertion – patella
   Action – abducts thigh and extends shank

4. **Biceps femoris** – very broad muscle posterior to the fascia lata.

   Origin – tuberosity of ischium
   Insertion – patella, tibia, and shank fascia
   Action – abducts thigh and flexes shank

5. **Tensor fasciae latae** – triangular mass of muscle anterior to the biceps femoris.

   Origin – ilium and fascia
   Insertion – fascia lata
   Action – tightens fascia lata

6. **Sartorius** – lateral part of this muscle is seen from the dorsal aspect of the thigh. Most of the muscle lies on the medial surface of the thigh.

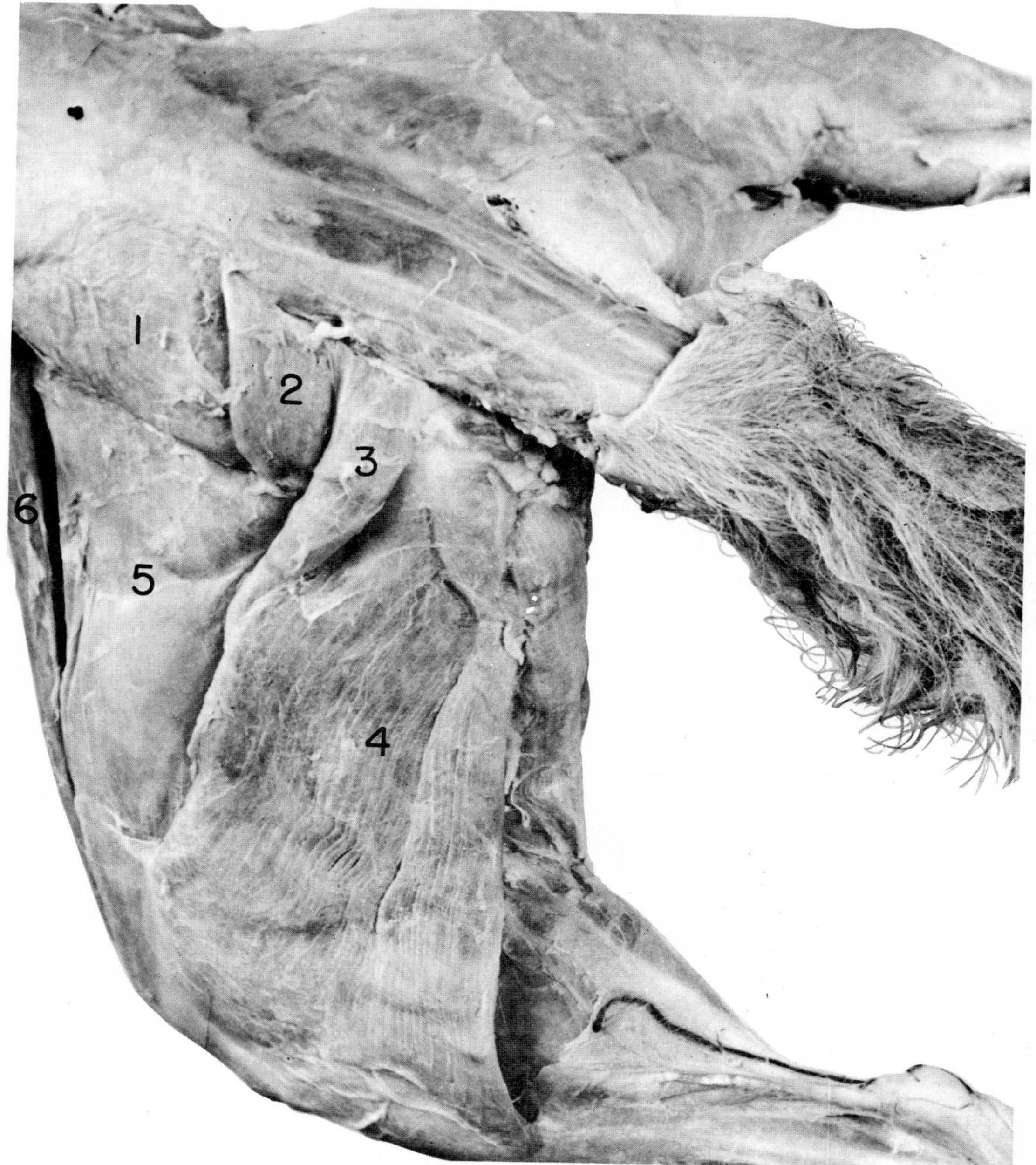

Fig. 31. Superficial muscles of the thigh, dorsal view.

1. Gluteus medius.
2. Gluteus maximus.
3. Caudofemoralis.
4. Biceps femoris.
5. Tensor fasciae latae.
6. Sartorius.

18

Deep Muscles of the Thigh

Dorsal Aspect (Fig. 32)

Transect the biceps femoris muscle and reflect. Cut across the fascia lata and reflect it with its tensor fasciae latae muscle. Clean the area of fat and tissue.

1. **Vastus lateralis** – this large muscle lies under the fascia lata and covers the anterolateral surface of the thigh.

   Origin – greater trochanter and shaft of the femur
   Insertion – patella
   Action – extends shank

2. **Tenuissimus** – a very slender band of muscle lying beneath the biceps femoris.

   Origin – second caudal vertebra
   Insertion – tibia (via fascia of the biceps femoris)
   Action – assists biceps femoris in abducting thigh and flexing shank

3. **Semitendinosus** – a large band of muscle covering the posteromedial border of the popliteal fossa.

   Origin – ischial tuberosity
   Insertion – medial side of tibia
   Action – flexes the shank

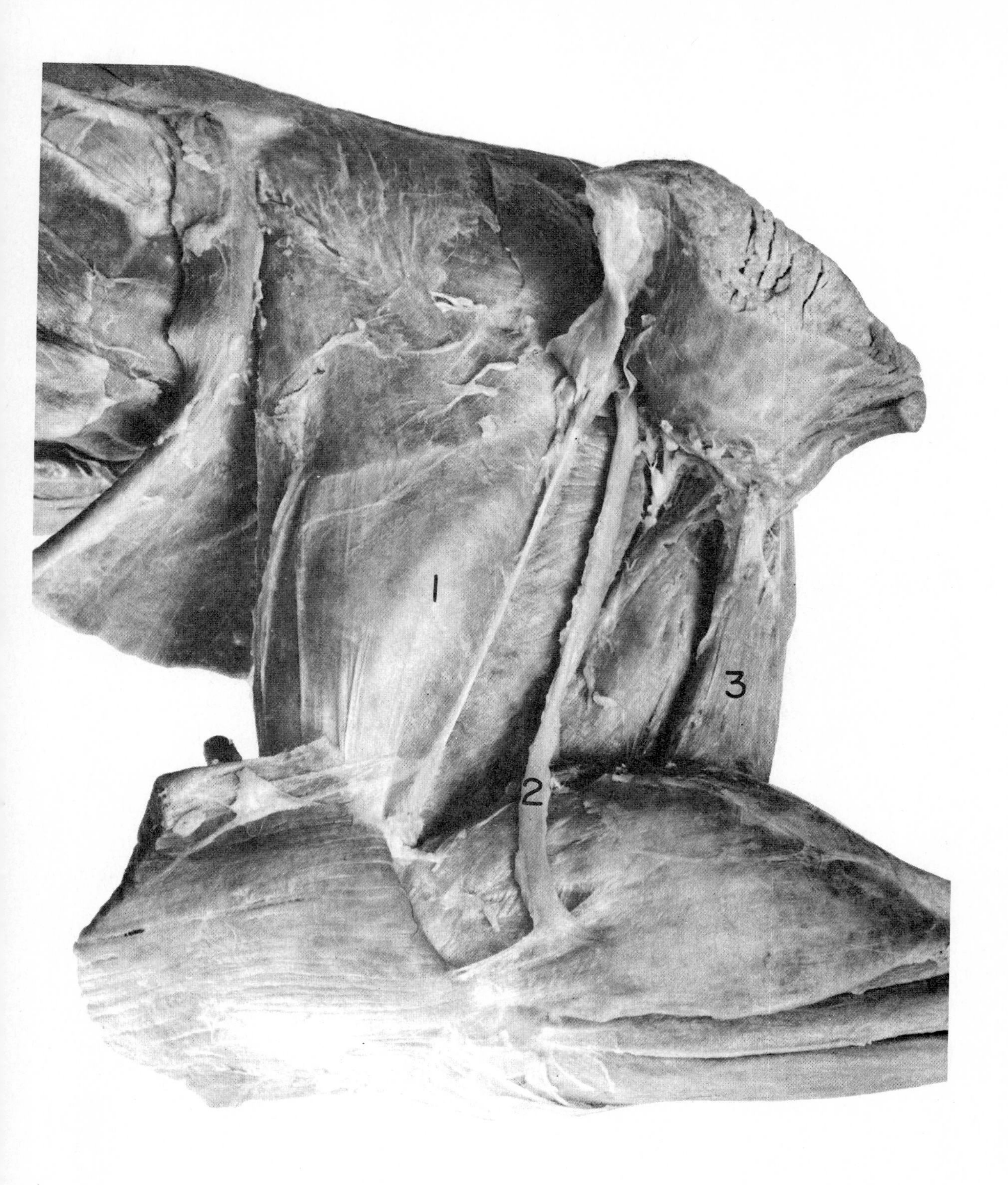

Fig. 32. Deep muscles of the thigh, dorsal aspect.
1. Vastus lateralis.
2. Tenuissimus.
3. Semitendinosus.

19

Deep Muscles of the Thigh

Medial Aspect – Ventral View (Fig. 33)

Cut across the belly of the sartorius and gracilis and reflect these muscles as far back as possible. The deeper muscles that are exposed separate very easily from one another.

1. **Rectus femoris** – narrow band of muscle on the front of thigh medial to the vastus lateralis. It lies beneath the sartorius.

   Origin – ilium
   Insertion – patella
   Action – extends shank

2. **Vastus medialis** – large muscle on medial surface of thigh, posterior to the rectus femoris. It is covered by the sartorius.

   Origin – shaft of the femur
   Insertion – patella
   Action – extends shank

Pull the rectus femoris and vastus lateralis widely apart. The deep muscle lying between them is the **vastus intermedius** (not seen in the illustration). It is difficult to separate this muscle from the **vastus medialis,** and the two may be continuous for part of their length. These four muscles form the four parts of the **quadriceps femoris** muscle (rectus femoris, vastus lateralis, vastus intermedius, and vastus medialis).

3. **Adductor longus** – a small, triangular muscle lying anterior to the adductor femoris.

   Origin – front of pubis
   Insertion – proximal end of femur
   Action – adducts thigh

4. **Adductor femoris** – large triangular mass of muscle posterior to the longus.

   Origin – ischial and pubic symphyses
   Insertion – shaft of the femur
   Action – adducts and extends thigh

5. **Semimembranosus** – large muscle posterior to the adductor femoris.

   Origin – posterior edge of ischium
   Insertion – medial epicondyle of femur
   Action – extends thigh

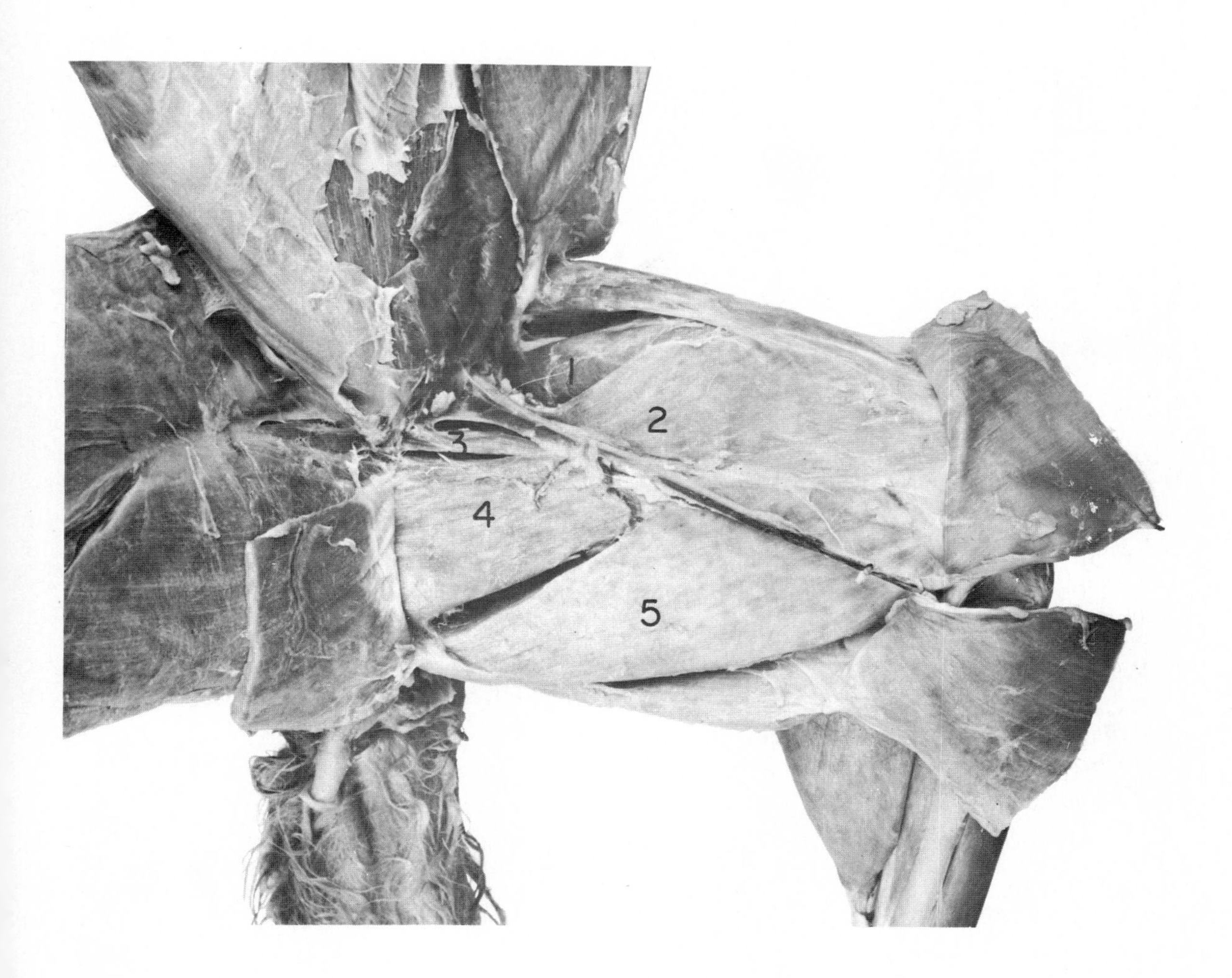

Fig. 33. Deep muscles of the thigh, medial aspect, ventral view.
1. Rectus femoris.
2. Vastus medialis.
3. Adductor longus.
4. Adductor femoris.
5. Semimembranosus.

20

Muscles of the Shank

Lateral Surface (Fig. 34)

Remove the tough fascia covering the shank and separate the obvious muscles.

1. **Gastrocnemius-lateral head** – this muscle forms the greater part of the calf of the leg. It has two heads, a lateral and medial; the medial head can be seen on a medial view of the shank.

   Origin – lateral condyle of femur
   Insertion – by a large tendon, the tendon of Achilles, on the calcaneus
   Action – extends foot, flexes shank

2. **Soleus** – small band of muscle lying next to the lateral head of the gastrocnemius.

   Origin – lateral surface of the head of the fibula
   Insertion – calcaneus by tendon of Achilles
   Action – extends foot

3. **Peroneus** – elongated band of muscle occupying the fibular aspect of the shank.

   Origin – lateral aspect of the head of the fibula
   Insertion – bases of the metatarsals and phalanges
   Action – extends foot and assists with plantar flexion. Prime mover for eversion of foot.

4. **Tibialis anterior** – a slender muscle situated on the anterolateral aspect of the tibia.

   Origin – proximal lateral surface of the tibia and corresponding portion of the interosseous membrane which joins the tibia and fibula.
   Insertion – base of first metatarsal
   Action – dorsiflexion and inversion of foot.

5. **Extensor digitorum longus** – slender, tapered muscle, which must be separated from the tibialis anterior, which overlaps it.

   Origin – lateral epicondyle of femur
   Insertion – phalanges
   Action – extends phalanges

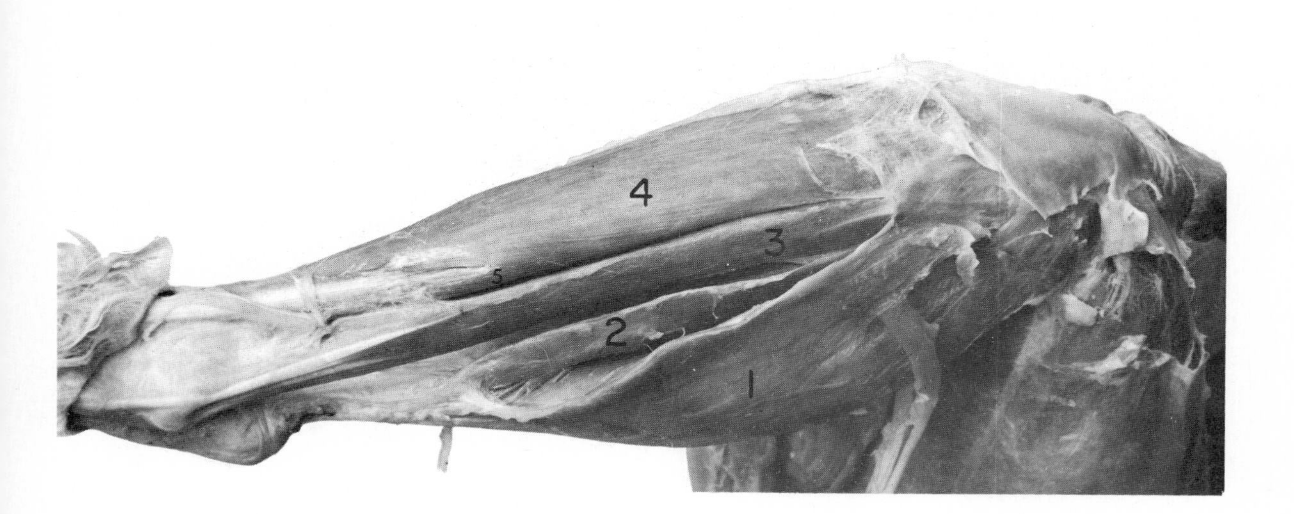

Fig. 34. Muscles of the shank, lateral view.
1. Gastrocnemius, lateral head.
2. Soleus.
3. Peroneus.
4. Tibialis anterior.
5. Extensor digitorum longus.

21

Muscles of the Shank

Medial Surface (Fig. 35)

1. **Gastrocnemius** – medial head

   Origin – medial condyle of femur
   Insertion – calcaneus
   Action – extends foot

2. **Flexor digitorum longus** – the first head (2a) is covered partly by the soleus and lies against the ventral surface of the tibia and fibula. The second head (2b) lies between the long head and medial aspect of the gastrocnemius.

   Origin – first head – tibia
   second head – fibula and fascia
   Insertion – in common by a tendon to the digits
   Action – flexes phalanges

3. **Tibialis anterior** – tapered band of muscle on anterior aspect of tibia.

   Origin – proximal ends of tibia and fibula
   Insertion – first metatarsal
   Action – flexes foot

4. **Plantaris** – the two heads of the gastrocnemius muscle meet behind in an indistinct line. If the two heads are separated, a strong round muscle will be revealed enclosed by the two heads. This muscle is the plantaris and is not shown in the illustration.

   Origin – patella
   Insertion – calcaneus and phalanges
   Action – flexes phalanges

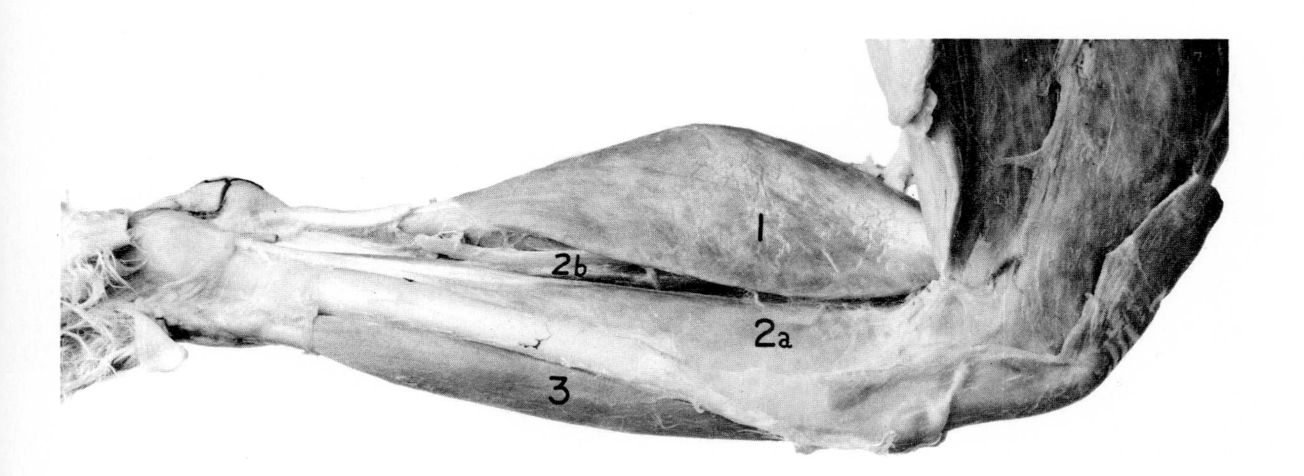

Fig. 35. Muscles of the shank, medial view.

1. Gastrocnemius, medial head.
2. Flexor digitorum longus.
   a. First head.
   b. Second head.
3. Tibialis anterior.

22

Muscles of the Neck and Throat

Ventral Aspect (Fig. 36)

Complete the skinning of the cat on the ventral side of the neck and throat as far back as the chin. Clean away the loose connective tissue and fat to expose the neck and throat muscles. Most specimens are somewhat mutilated in this region, because the large vessels are used for preparation of the cat in the preserving process.

1. **Sternomastoid** – paired large bands of muscle covering front of the throat extending dorsally toward the mastoid region of the skull.

   Origin – mastoid region of skull
   Insertion – manubrium
   Action – singly, turns head; acting together, depress head on neck

2. **Cleidomastoid** – narrow band of muscle lateral to the sternomastoid, running dorsally to the clavotrapezius and clavodeltoid.

   Origin – mastoid process
   Insertion – clavicle
   Action – turns and lowers head

3. **Clavotrapezius and clavodeltoid** – a large band of muscle located on the back side of the neck dorsal and lateral to the cleidomastoid. The most anterior part is the clavotrapezius, which arises from the back of the skull. The clavodeltoid continues from the insertion of the clavotrapezius to the arm. These two muscles have united and are sometimes described as one, called the **cephalobrachial** or **cephalohumeralis.**

   Clavotrapezius

   Origin – back of skull and spinous process of first four cervical vertebrae
   Insertion – clavicle
   Action – draws clavicle dorsally and toward the head (cranially)

Clavodeltoid

Origin – clavicle
Insertion – ulna
Action – flexes forearm

4. **Sternohyoid** – by pushing the sternomastoid muscles laterally, a thin band of muscle lying directly in the midventral plane and covering the windpipe will be exposed. This is the sternohyoid, which is actually a pair of muscles that have fused.

   Origin – sternum
   Insertion – hyoid bone
   Action – draws hyoid posteriorly

5. **Sternothyroid** – a narrow band of muscle lying lateral and dorsal to the sternohyoid. To be seen, the sternohyoid must be lifted and pushed somewhat aside.

   Origin – sternum
   Insertion – thyroid cartilage of larynx
   Action – pulls larynx posteriorly

6. **Thyrohyoid** – a narrow band of muscle located just above the sternothyroid. These two muscles appear as one band unless their attachments to the larynx are carefully exposed.

   Origin – thyroid cartilage of larynx
   Insertion – hyoid
   Action – raises the larynx

7. **Geniohyoid** – a midventral band of muscle located anteriorly to the sternohyoid. The mylohyoid must be cut and reflected to see this muscle.

   Origin – mandible
   Insertion – hyoid
   Action – draws hyoid forward (anteriorly)

8. **Mylohyoid** – sheet of muscle with transverse fibers lying deep between the digastric muscles. The digastric muscles described below must be separated to expose the mylohyoid.

   Origin – mandible
   Insertion – hyoid
   Action – raises floor of mouth, pulls hyoid forward (anteriorly)

9. **Digastric** – stout band of muscle attached to the ventral border of the lower jaw (mandible).

   Origin – occipital and mastoid process of skull
   Insertion – ventral border of lower jaw (mandible)
   Action – opens jaw

10. **Masseter** – large, round, powerful muscle in the cheek, lying ventral to the zygomatic arch.

    Origin – zygomatic arch
    Insertion – coronoid process of mandible
    Action – elevates jaw

11. **Temporalis** – this large muscle lies dorsal to the zygomatic arch and fills the large temporal fossa. This muscle may be exposed by making a longitudinal incision on the scalp from the back of the nose to the occipital region on the skull. Reflect the skin laterally. Two large, fleshy muscles lying in the temporal fossa will be exposed. This muscle is not shown in the illustration.

Origin – superior nuchal line
Insertion – coronoid process of mandible
Action – elevates jaw with masseter

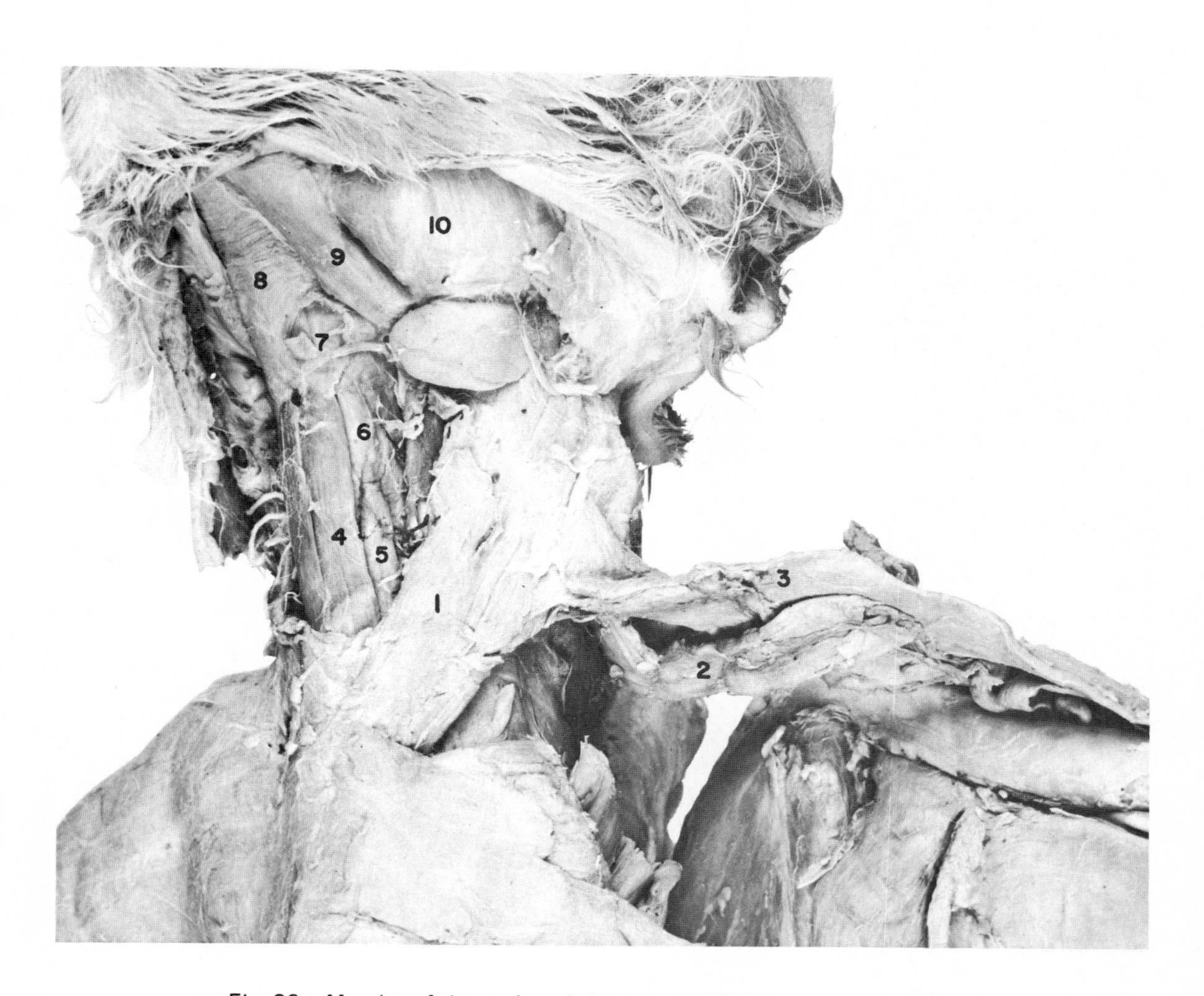

Fig. 36. Muscles of the neck and throat, ventral view.

1. Sternomastoid.
2. Cleidomastoid.
3. Clavotrapezius and clavodeltoid.
4. Sternohyoid.
5. Sternothyroid.
6. Thyrohyoid.
7. Geniohyoid.
8. Mylohyoid.
9. Digastric.
10. Masseter.

23

Muscles of the Face

See Fig. 37.

Epicranius – draws scalp back

Corrugator supercilii – draws eyebrows downward and medially

Orbicularis oculi – closes eyelids, wrinkles forehead, compresses lacrimal sac

Orbicularis oris – closes lips

Buccinator – pulling lips against teeth aids in mastication, swallowing and whistling

Platysma – depression of mandible, depression of lower lip, wrinkling skin of neck

Masseter – elevation of jaw

Temporal – elevation of jaw, retraction and rotation of jaw

External pterygoid – brings jaw forward

Internal pterygoid – raises mandible, closes mouth

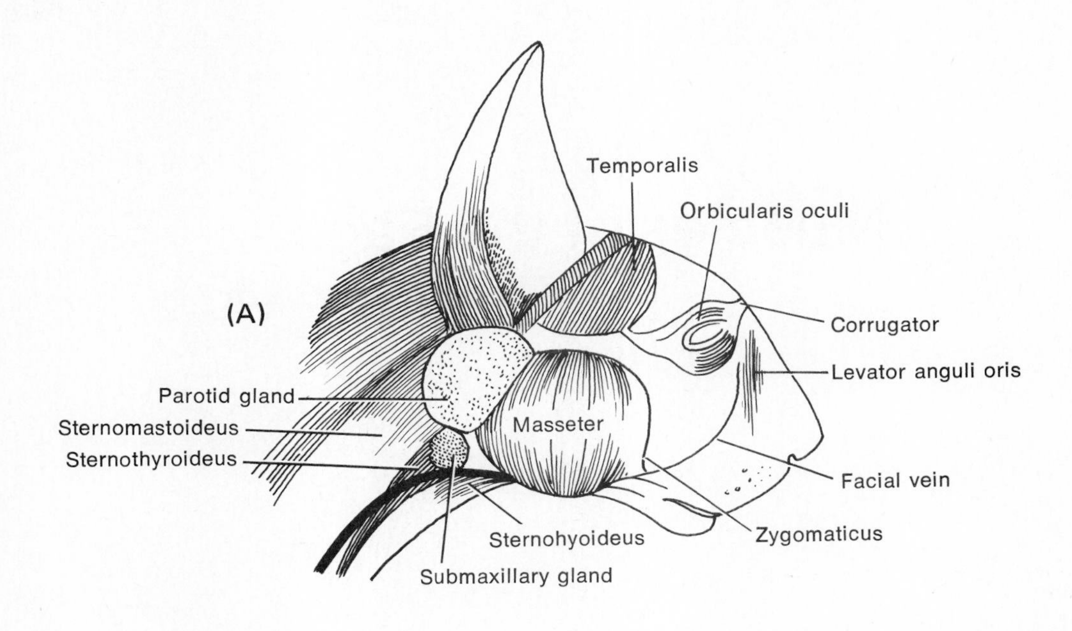

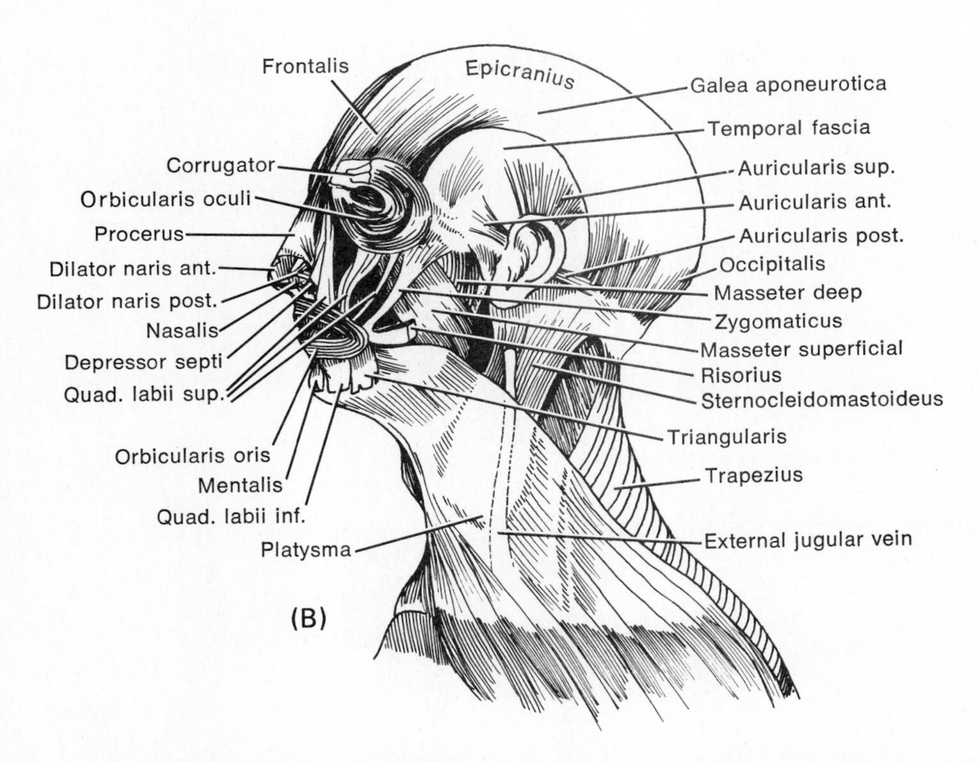

Fig. 37. The major head and neck muscles of the cat contrasted with the major head and neck muscles of the human.

24

Muscles of the Eye

Superior rectus – rolls eyeball upward

Inferior rectus – rolls eyeball downward

Lateral rectus – rolls eyeball laterally

Medial rectus – rolls eyeball medially

Superior oblique – rotates eyeball on axis; directs cornea downward and laterally

Inferior oblique – rotates eyeball on its axis; directs cornea upward and laterally

Levator palpabrae superioris – raises upper lid

Fig. 38. Anterior and posterior view of human showing major muscles.

1. Frontalis M. 2. Orbicularis oculi M. 3. Masseter M. 4. Orbicularis oris M. 5. Sternomastoid M. 6. Trapezius M. 7. Pectoralis maj. M. 8. Deltoid M. 9. Biceps M. 10. Serratus ant. M. 11. Sheath of rectus abdominis. 12. Linea alba. 13. Ext. oblique M. 14. Brachioradialis M. 15. Flex. carpi rad. M. 16. Palmaris long. M. 17. Palmar aponeurosis 18. Gluteus med. M. 19. Iliopsoas M. 20. Pectineus M. 21. Adductor long. M. 22. Gracilis M. 23. Sartorius M. 24. Rectus M. 25. Vastus lat. M. 26. Vastus medialis M. 27. Patella. 28. Ant. tibial M. 29. Gastrocnemius M. 30. Teres maj. M. 31. Triceps M. 32. Latissimus dorsi M. 33. Flexor carpi ulnaris M. 34. Extensor carpi ulnaris M. 35. Extensor dig. V. M. 36. Extensor dig. com. M. 37. Iliac crest. 38. Glut. max. M. 39. Iliotibial tract. 40. Biceps femoris M. 41. Semitendinosus M. 42. Semimembranosus M. 43. Calcaneal tendon (Achilles).

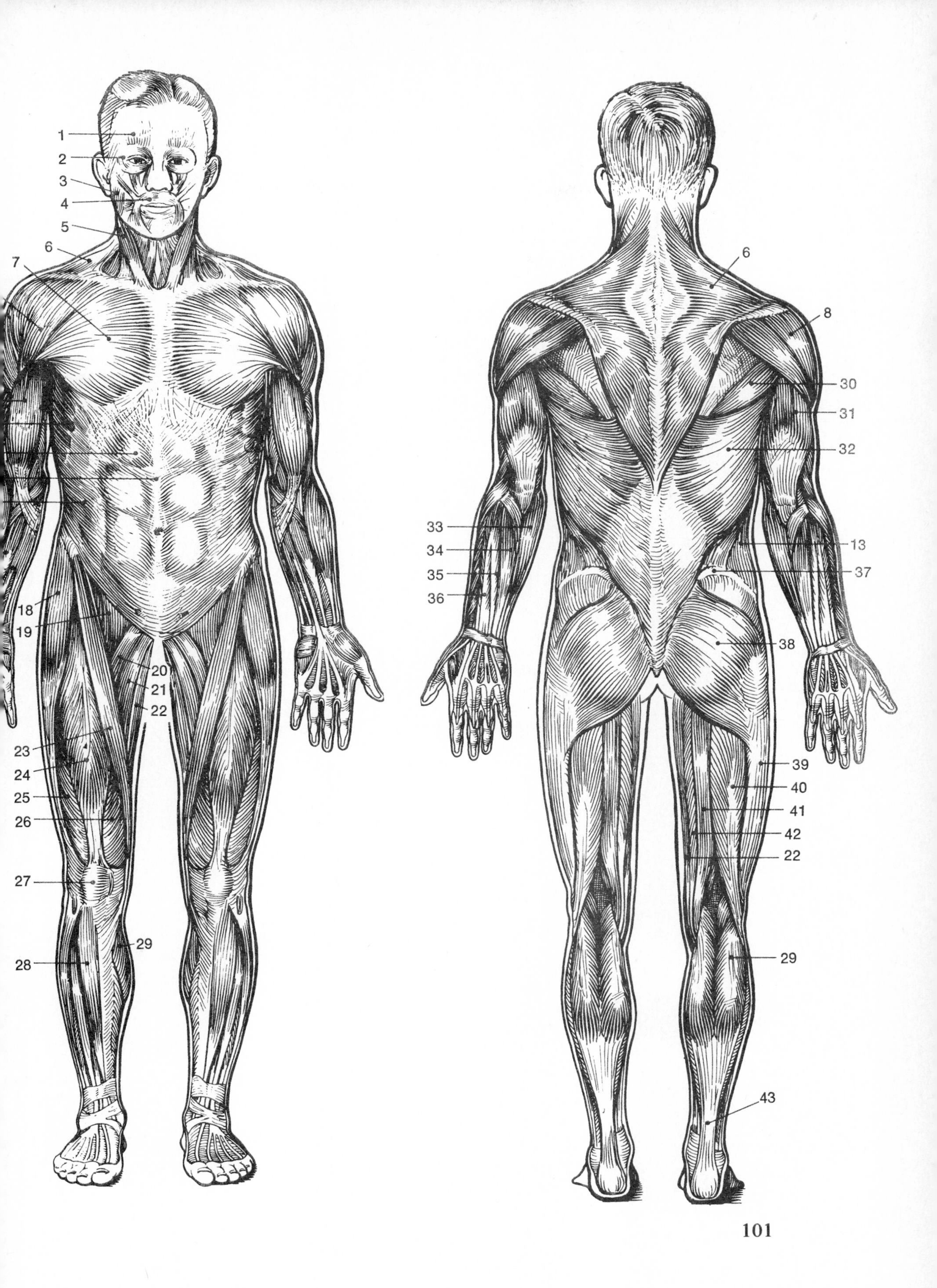
1
2
3
4
5
6
7
18
19
20
21
22
23
24
25
26
27
28
29
6
8
30
31
32
33
34
35
36
13
37
38
39
40
41
42
22
29
43

25

Apparatus Used in Physiological Experiments

Apparatus

KYMOGRAPH

The function of a kymograph (Fig. 39A) is to record motion. The three important parts of any kymograph are

1. A surface upon which to record motion
2. A mechanism for moving this surface at a uniform rate
3. A mechanism for changing speed

In early kymographs the surface was that of a cylindrical drum moved electrically or by means of clockwork. Today, horizontal chart movers (Fig. 40A) are used in most instances.

Study the following (your laboratory instructor will demonstrate if your laboratory is not equipped with electrically driven ink recorders):

1. How to wind the clockwork
2. How to regulate the speed of the drum
3. How to lower and raise the drum on its axis
4. How to mount and smoke the paper on the drum
5. How to remove the paper by cutting
6. How to label your tracings; each tracing must be labeled as follows:
   a. Proper legend, stating the nature of the curve
   b. Proper lettering (a, b, c, and so on) in the individual curves showing the characteristic parts
   c. Your name or names of all students in a group if it is a group tracing and date
7. How to varnish and dry the tracing if smoked drum is being used

TUNING FORK (Fig. 39F)

To study the duration of the various phases of a certain movement, it is necessary at the time that the movement itself is recorded to produce a record by which time can be measured. A method frequently employed is that of recording the vibrations of a tuning fork. The Harvard tuning fork makes 100 double vibrations per second. An electrically driven tuning fork is much easier to use.

SIGNAL MAGNET OR TIME MARKER

The divisions of time recorded by the tuning fork are too small for certain purposes. To read seconds and fractions of minutes, the signal magnet (Fig. 39J) is employed. The signal magnet is also used for indicating on the kymograph any change in the procedure, such as the time of stimulation or the application of a chemical. In the standard event/time marker module (Fig. 40E) the pen is deflected to mark intervals of 1 sec. Each 10 sec are marked by a higher amplitude line, and each minute with still a higher line. Contained within the event/time marker module is a signal magnet that deflects the marker pen in the opposite direction from that of the time base when a stimulus is applied.

INDUCTORIUM OR INDUCTION STIMULATOR

The important parts of the inductorium are a primary coil and a larger secondary coil. There is no metallic connection between the wires of one coil and those of the other. A galvanic current traveling through the primary coil will induce a current in the secondary coil. This is known as **induced current**, or **faradic current**, and is the type of electrical current used as a stimulus in most physiological experiments.

The typical **sliding coil type inductorium** (Fig. 39D) when used as a source of stimulation voltage, is run on one or two dry cells. It is able to furnish simple make and break stimuli or tetanizing currents of about 50 cycles per second. The voltage may be adjusted by the position of the secondary coil from 0 to 2400 V. The **induction stimulator** (Fig. 40C) is similar in function to the inductorium and consists of batteries, a vibrating transformer, and controls enclosed in a compact case. A stimulation voltage from 0 to 500 V can be selected by a range switch. A lever switch selects either make and break operation or tetanizing stimuli at 50 cycles per second.

BATTERIES. An electrical stimulus is usually employed in physiological experiments for the following reasons:

1. It leaves the tissue relatively unharmed
2. It is more easily controlled
3. It can be more accurately measured

The batteries to supply this stimulus are usually 1½ V with a positive and a negative pole.

Key. Simple contact key to make or break the electrical circuit.

PLATINUM ELECTRODES. Used to stimulate specific areas of tissue under study; bare copper wires that do not touch one another may be used.

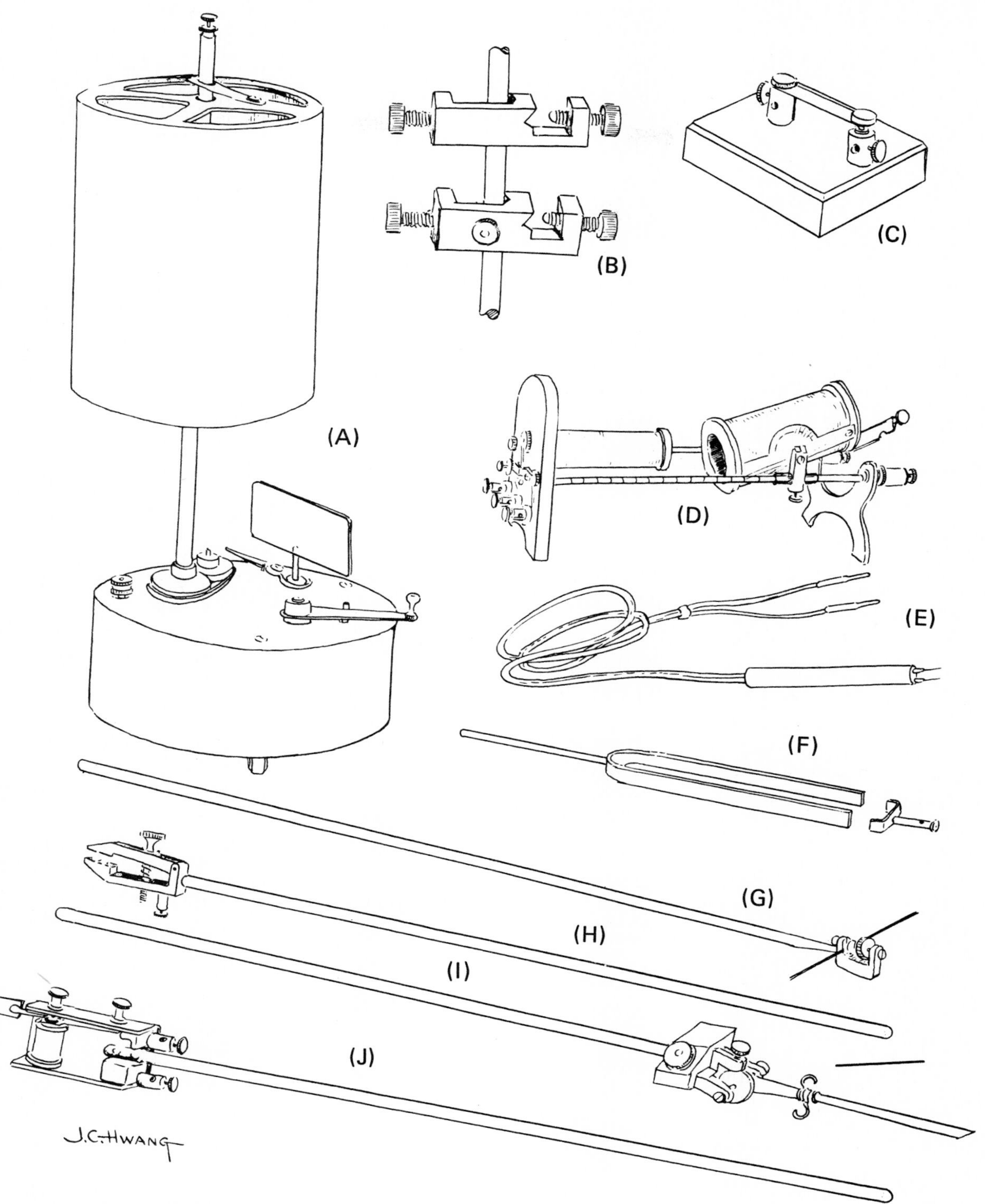

Fig. 39. Physiological apparatus.

A. Spring kymograph and drum.
B. Double clamp.
C. Simple key.
D. Inductorium.
E. Platinum electrode.
F. Tuning fork and tuning fork starter.
G. Heart Lever.
H. Femur clamp.
I. Muscle lever.
J. Signal magnet.

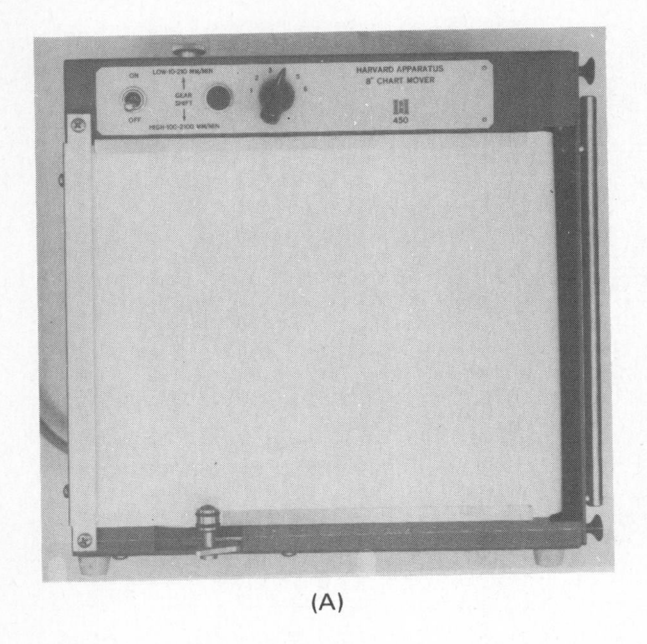

(A)

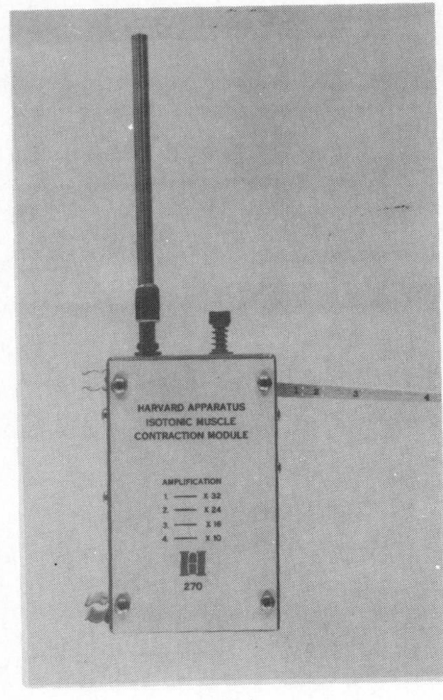

(B)

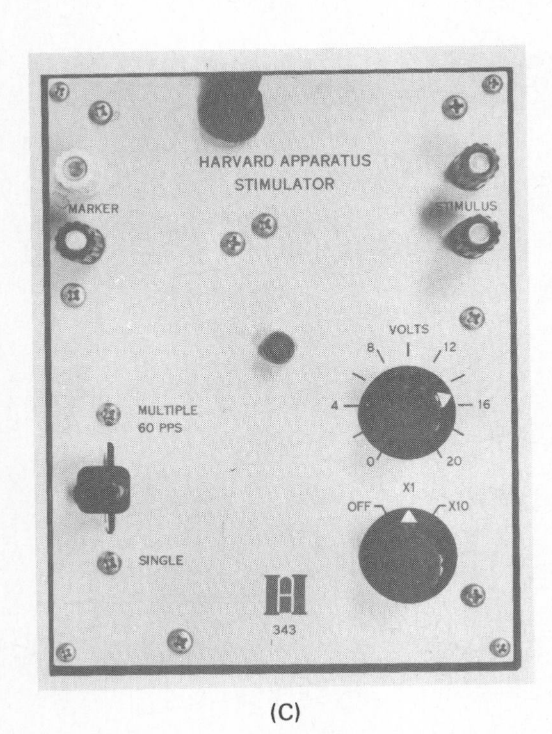

(C)

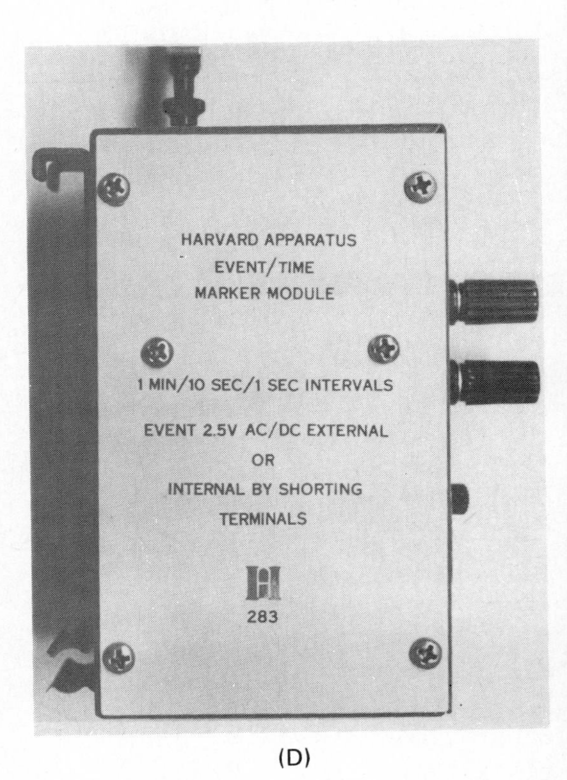

(D)

Fig. 40. Physiological mechanical transducer modules.
A. Chart mover.
B. Isotonic muscle contraction module.
C. Induction stimulator.
D. Event/time marker module.

MUSCLE CLAMP. To hold femur of the muscle preparation.

MUSCLE LEVER OR ISOTONIC MUSCLE CONTRACTION MODULE

Muscle lever (Fig. 39I) – consists of the following parts:

1. Pen – to record muscle contraction
2. Axis with a wheel – for attachment of muscle
3. Afterloader – to relieve stretch on muscle by added weight
4. Set screw for attachment of wire

Muscle Contraction Module (Fig. 40B) consists of the following parts packaged in a compact unit:

1. Ink writing pen – to record muscle contraction
2. Stainless steel rod – for attachment of femur clamp
3. Aluminum lever bar, calibrated in four positions for different amplifications: 10X, 16X, 24X, 32X – the bar serves for both the attachment of the gastrocnemius muscle and the weighted scale pan.
4. Afterload control – to relieve stretch on muscle by added weight.

The various muscle modules are so constructed that motion of a lever in a vertical plane is translated to rotary horizontal motion and the corresponding movement of an ink writing pen.

All apparatus listed may be ordered from the Harvard Apparatus Company, Inc. 150 Dover Road, Millis, Massachusetts, 02054.

Physiological Solutions

The following solutions are use for cold-blooded (frog) and for warm-blooded (mammal) animals.

FROG RINGER'S (isotonic solution)

Sodium chloride	6.0 g
Potassium chloride	0.14 g
Calcium chloride	0.12 g
Sodium bicarbonate	0.20 g

Distilled water sufficient to make 1000 ml (1 ℓ)

MAMMALIAN RINGER'S (isotonic solution)

Sodium chloride	8.5 g
Potassium chloride	0.42 g
Calcium chloride	0.24 g
Glucose	1.00 g

Distilled water sufficient to make 1000 ml (1 ℓ)

26

Electronic Recording Devices

Multichannel recording devices are being used more extensively in undergraduate teaching laboratories than ever before. Electronic recording devices provide the student with an instrument for recording physiological quantities with increased accuracy and less time than required with the classical kymograph setup.

The principle underlying all the electronic recording devices on the market is the same, since it is based on the recording of information (physiological data) that changes with time. There are three basic components that make up the electronic recorder: the sensing device, called the **transducer**, the amplifying device, called the **processer**, and the recording device (an ink writing module), called the **reproducer**, which converts the information into the form of a graphic record.

TRANSDUCER

Transducers convert the physiological event into a proportional electrical signal. They do this through changes in inductance, resistance, capacitance, or voltage. The most common sensing device in transducers is a light bulb and photoelectric cell. The sensitivity is controlled by the brightness of the bulb, and the electrical signal is generated by the photoelectric cell when light from the bulb strikes it, causing the photocell to emit electrons. The number of electrons flowing will be proportional to the amount of light striking the photoelectric cell. A shutter is interposed between the light source and phototube, and movement of the shutter is usually brought about by some mechanical force, such as a muscle contraction that allows light to strike the phototube. The greater the force of contraction, the greater the movement of the shutter away from screening the light source, and the greater the emission of electrons from the phototube. The block diagram on the following page illustrates the principle of photoelectric force transducers.

Myographs are transducers used primarily for quantitative measurements of smooth, cardiac, and skeletal muscle contractions. Muscle movement is transmitted by a loop of thread attached to a hook on the myograph shutter. **Bellows pneumographs** are photoelectric force transducers for recording respiration. The **blood pressure transducer** is a photoelectric transducer for

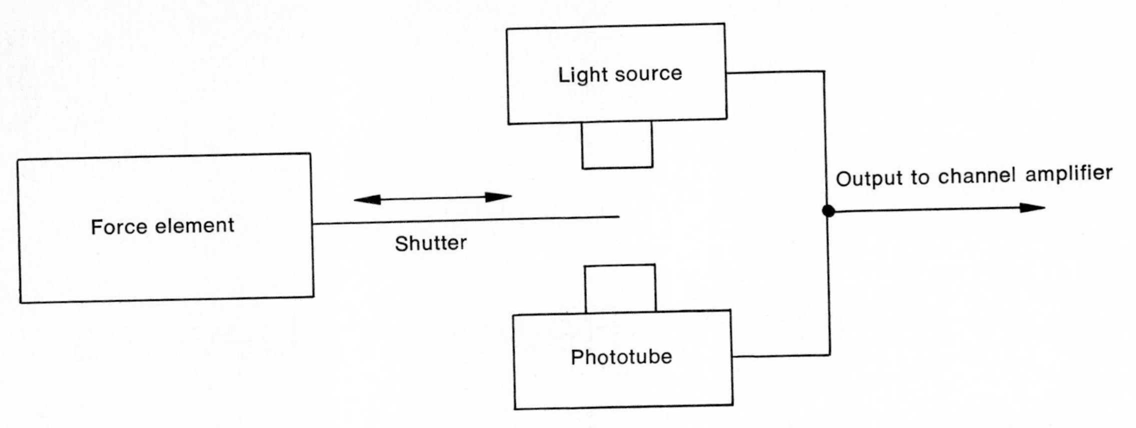

quantitative measurement of liquid or gas pressures. It is normally used for measurement of direct blood pressures where cannulation is possible.

In recording the electrical activity of the heart (electrocardiogram), transducers are not necessary, as the electrical activity may be picked up directly from the surface of the body by plate electrodes and led into an amplifier, which enlarges it. After amplification, the electrical signal is powerful enough to drive the recording galvanometer, which converts the electrical signals into movements of a pen, allowing them to be recorded on paper.

AMPLIFIER

Since the output signal from transducers is small, it must be amplified considerably in order to drive the recording pen motor. The function of the amplifier is to provide the required increase in signal strength, while retaining the proportional relationship of the electrical signal of the transduced physiological activity. The current put out by the amplifier drives the pen motor or reproducer.

REPRODUCER

The reproducer is a recording device that converts electrical energy into rotary motion. It is usually a motor driven pen that receives the electrical signals from the amplifier and drives the pen across a moving chart, providing a record of the event being measured. Fig. 41 represents a typical recording channel.

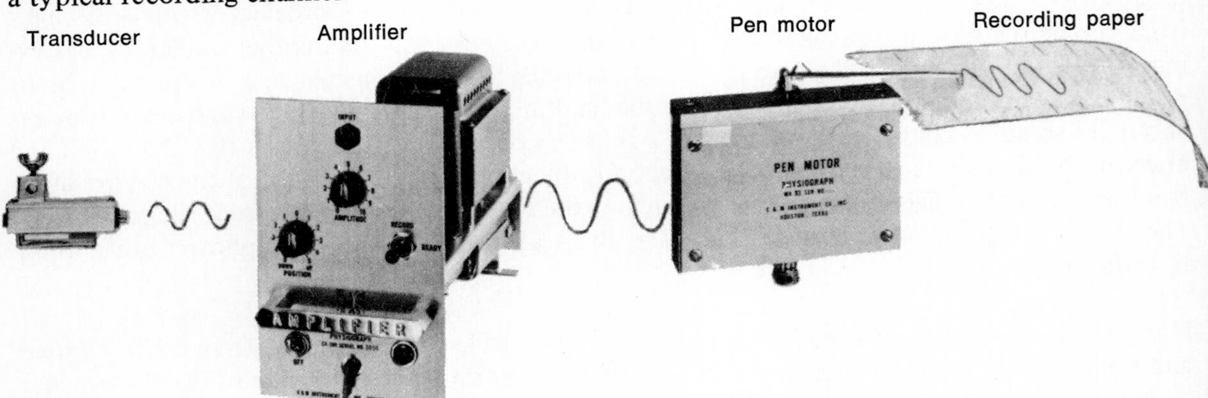

Fig. 41. Electronic recording channel.

27

Preparation and Handling of Materials

A. In the physiologic laboratory you are learning to deal with living tissues and organs. Success in your work depends very largely on the vitality of the organs you work with. The more this vitality is lost, the poorer are the experimental results. Because the tissues and organs are frequently excised from the body, or at least exposed, they are placed in an "unnatural" environment and deteriorate quickly. Retention of their normal properties is greatly aided by careful treatment on your part. It will be to your advantage, therefore, to read attentively the following suggestions.

1. Before starting the work, read the whole experiment. Be sure you understand the nature of the experiment and the results to be obtained. Ten or fifteen minutes spent in planning the procedure is time gained.
2. Never handle excised or exposed tissues or organs with forceps; use a glass hook or seeker or a small camel's hair brush moistened with Ringer's solution. Physical or chemical injury to the living structure is likely to ruin your experiment. When it is necessary to lay down an excised muscle, nerve, or heart, place it in a clean watch glass or similar dish moistened with Ringer's solution. *Never place an excised structure on the desk.*
3. Do not let the skin of the frog come in contact with the muscles or nerves; the secretion of the skin is poisonous to them.
4. Always keep excised or exposed tissues moist with Ringer's solution. Drying speedily kills them. Apply the Ringer's solution with a clean medicine dropper or a camel's hair brush.
5. Do not put your fingers on an excised or exposed tissue unless specifically told to do so. In the laboratory your fingers are not likely to be overclean.
6. Never unnecessarily "practice" with the living structure before beginning the experiment.
7. As soon as the tissue or organ has been prepared, begin the work. Do not lose any time – the tissue is already dying. The living preparation must never be ready before all the apparatus is set up and the problem is clearly understood. This point cannot be emphasized too strongly.

B. In kymograph work two or three students usually work together as a group. Let one student set up the recording apparatus, the second arrange the stimulating apparatus, and the third prepare the muscle or whatever tissue or organ may be the subject of experimentation.

Arrange the apparatus neatly and securely on the desk; use sufficiently long wires so that undesirable crowding is avoided.

C. To prepare a frog gastrocnemius muscle (with the femur and tendon of Achilles attached):

1. With a pair of large shears decerebrate or pith a frog. Cut away the trunk just above the femur bones. In a median line cut between the two hind legs. Remove the skin from the whole preparation by carefully and quickly peeling the skin, using forceps or fingers.
2. Lay the preparation on a clean plate moistened with Ringer's solution and remove the thigh muscle. Using scissors, cut the femur so that a short piece remains attached to the gastrocnemius.
3. Place one blade of the scissors between the tendon of Achilles and the gastrocnemius muscle and slip it down below the heel, loosening the tendon. Cut the tendon loose from its attachment below the heel so that a relatively large piece of tendon will be attached to the muscle.
4. Lift the tendon with forceps, and with scissors sever the fascia binding the gastrocnemius to the neighboring structures. The tibia-fibula now must be cut below the knee.
5. Keep tissue moist with frog Ringer's solution at all times. When attached to apparatus, use medicine dropper to wash muscle with Ringer's.

Experiment I

Observation of Normal Animal; Producing a Decerebrate Animal and a Spinal Animal

Reflex Activity

Although the frog is an amphibian, it has been used to study many of the basic physiological phenomena associated with mammalian forms. It is certainly a useful animal in this respect since it is relatively easy to obtain and maintain and is not as expensive as other forms of experimental animals. In addition, since it is a cold blooded animal (poikilotherm), it is not necessary to maintain the tissues being studied at constant temperatures, because the animal can adjust to normal fluctuation of environmental temperatures and still continue to function without abnormal reactions.

OBSERVATIONS OF NORMAL ANIMAL

1. Place frog on board or plate and observe normal posture.
2. Place the animal on its back and observe its righting movements.
3. Again place animal on its back and while holding it in this position note heart movement and attempt to determine rate.
4. Place the animal in a large beaker of water that is at room temperature and observe its swimming movements.
5. While the frog is in the beaker, determine its respiration rate by counting the pulsations in the throat area. These pulsations correspond to the inhalation-exhalation sequence and represent a direct indication of the respiration rate.
6. Touch the cornea of the eye with finger or dissecting probe.

DECEREBRATION OF ANIMAL

The cerebral hemispheres of the frog are very easily removed; simply cut off the entire front of the head. Open the mouth and with the very sharp blades of a scissors aligned across the upper jaw behind the eyes and in front of the tympanic membrane, cut off the forebrain by a single cut. This procedure removes all but the midbrain and medulla. After allowing several minutes for recovery from operative shock, repeat the observation carried out on the normal animal, except step 6.

THE SPINAL ANIMAL

All the brain may be destroyed by a single pith procedure. Holding the frog in one hand with the tip of its mouth resting between the middle and index finger, flex the head ventrally so that it forms right angles with the body; at the same time run the fingernail of the thumb of the other hand down the back of the head until you feel a depression.

This depression marks the junction between the skull and the vertebral column. Carefully insert a dissecting needle by its sharp end through the skin at a point in the midline of the depression; thrusting the needle downward and forward should take it through the opening of the magnum foramen into the cranial cavity.

By twisting the needle vigorously, all the brain can be destroyed without injuries to the spinal cord. The laboratory instructor will demonstrate. Such an animal is referred to as a **spinal animal.** After waiting several minutes to allow for some recovery from shock, again observe the activities listed above and record your findings.

Suspend the spinal animal on a stand with a flat jawed clamp attached to its lower jaw and make observations of the following reflexes.

Reflex	*Present*	*Not Present*
Muscle tone: Are the limbs completely flaccid or is there some tension?		
Flexion response: Pinch the toes of one leg with forceps. Does the leg flex?		
Crossed extension reflex: Make stimulation extremely strong. As leg flexes, the opposite limb extends.		
Purposeful activity: Place some irritating substance such as acetic acid on the surface of the belly. A purposeful type of leg reaction toward irritated spot may occur.		

Remove the spinal animal from the stand and destroy the spinal cord. This procedure is carried out in the same manner as the pithing except that the needle is directed downward into the spinal column. This preparation is referred to as a **doubly pithed frog**. After the animal has recovered from the spinal shock, again make observations of the presence of spinal reflexes. Discuss your results.

Experiment II

Muscle Twitch—Simple Muscle Contraction

APPARATUS

Stimulator (dry cells, simple key, induction coil, wire), kymograph, several stands, femur clamp, muscle lever with pen or muscle contraction module, and signal magnet or time/event module. Glass plate or frog board, Ringer's solution, thread, and 10-g weight.

PROCEDURE

1. Smoke drum. Fasten muscle clamp to stand and below it a muscle lever. Set up the stimulating apparatus and recording apparatus from instructions given by laboratory instructor. In the metal ends of the lever place a metal writing pen to the end of which affix a stylus (small piece of celluloid attached to pen by small wad of beeswax). See Figs. 42 and 43 for proper wiring and apparatus arrangement.

2. Insert femur of muscle preparation in the jaws of the muscle clamp. Attach tendon end to muscle lever with a piece of thread. Suspend a 10-g weight just below the muscle lever where the muscle is attached, making sure the resting position of the muscle lever is horizontal. When the muscle is stimulated, it will contract and then relax. The pen will move up and down when this occurs. To obtain perfect recordings of this activity, the penpoint must slide freely over the smoked paper without excess friction. This result is not obtained unless the pen moves in the vertical plane of the drum, and for this to take place the lever must be at a right angle to the vertical plane of the drum.

3. Use a medium speed drum. Use about 1½ V in the primary circuit. Close the key for a few seconds and then open the key. With a sufficiently strong current a contraction occurs at both the closing and opening of the key. Label these "M" and "B" (make and break), respectively. If the current is too strong, several twitches will result. Label tracing "Muscle twitch, faradic stimulation." Make a tracing for each individual member of group.

4. Fix the tracing by carefully removing the smoked paper from the drum and coating the paper with shellac.

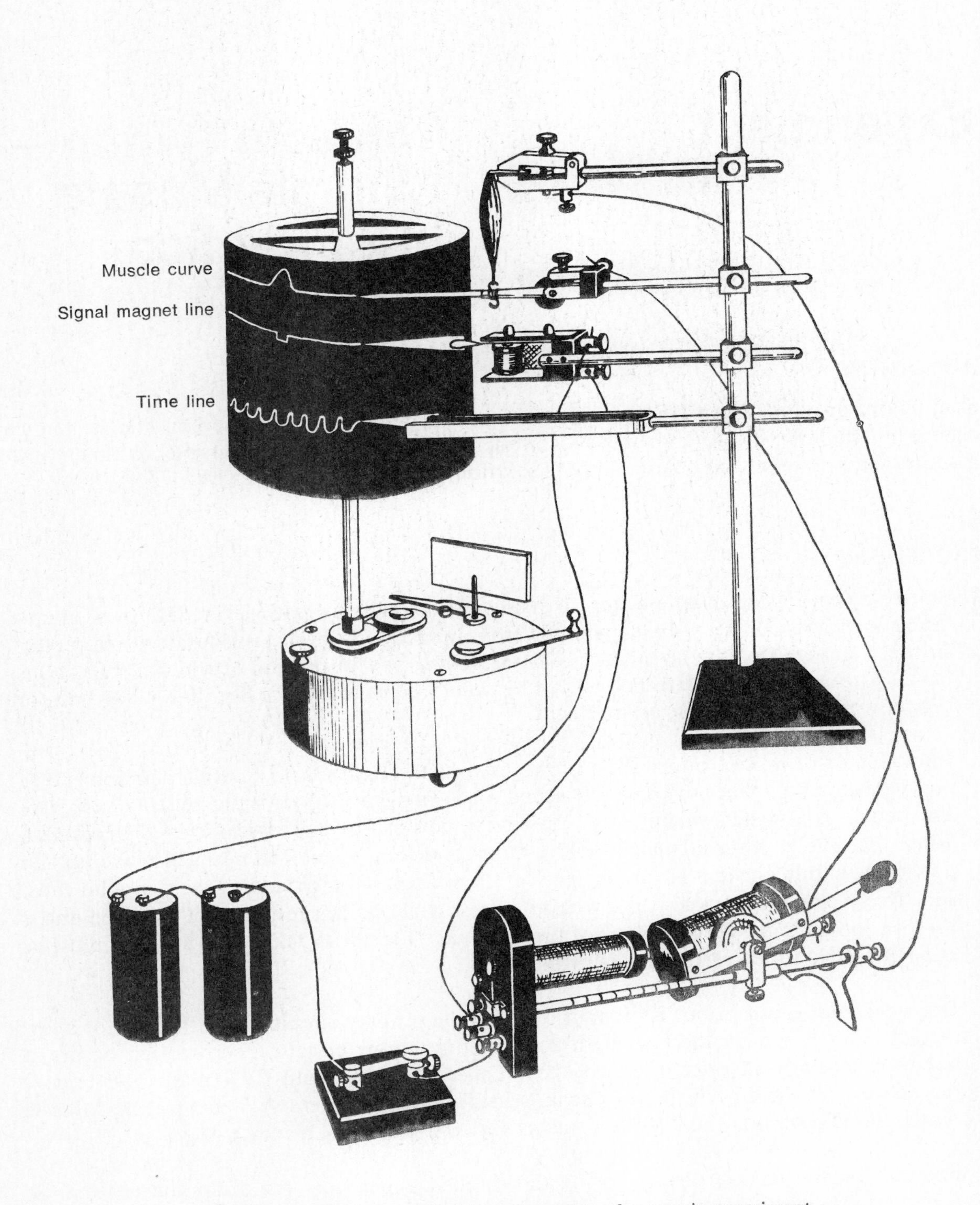

Fig. 42. Arrangement and wiring of apparatus for muscle experiment.

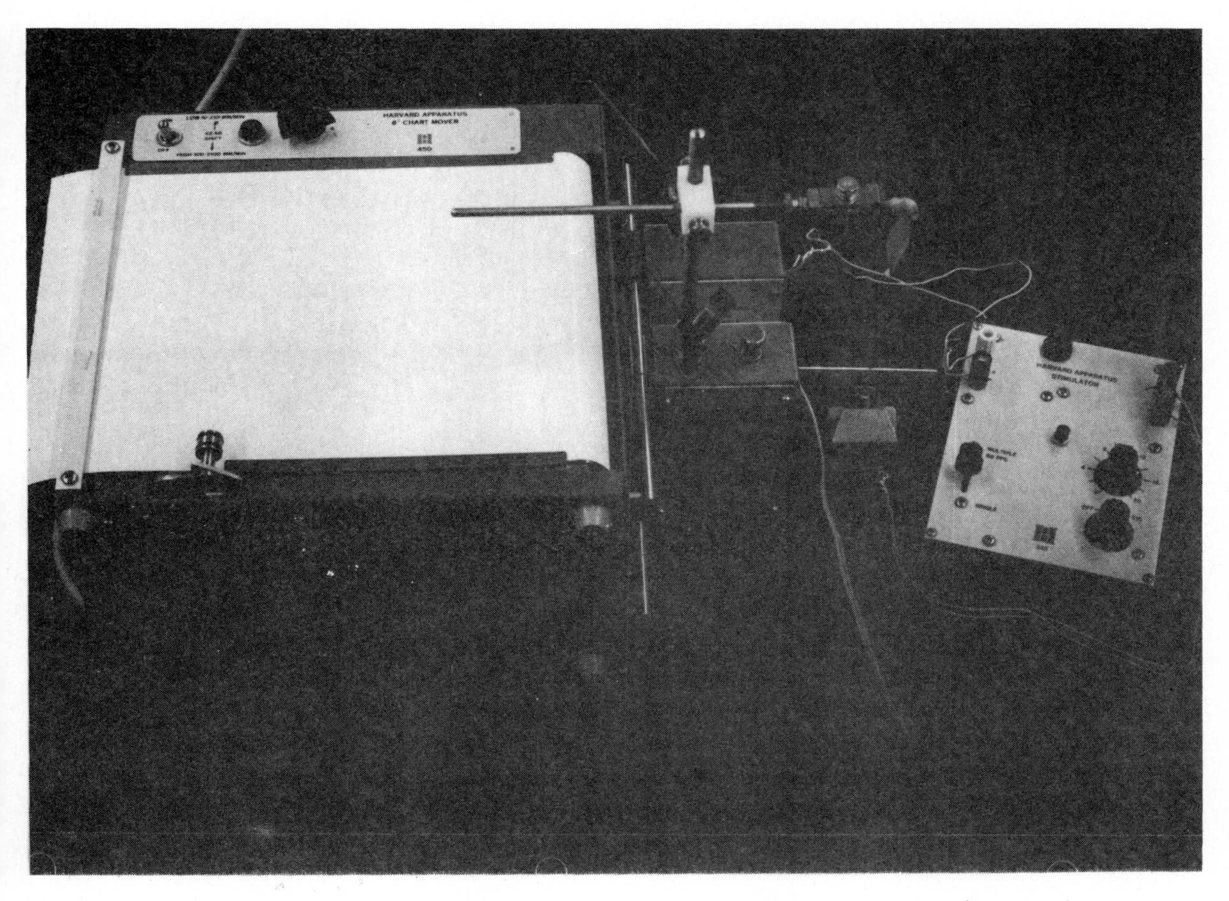

Fig. 43. Typical set up for muscle experiment using the chart mover, event/time marker and isotonic muscle contraction module (Harvard Apparatus, Millis, Ma.)

Experiment III

The Effect of Variation of Strength of Stimulus and Height of Muscle Contraction

APPARATUS

Same as in muscle-twitch experiment.

PROCEDURE

1. In all tracings let the resting muscle write a short introductory line on the drum (base line) if a drum kymograph is being used. This is not necessary if a chart mover is being used. Set up the apparatus as in the experiment on muscle twitch (tuning fork not necessary). For easy calculation, construct the lever so that the distance from the axis to the point of the pen is a simple multiple of the distance from the axis to the point of attachment of the muscle on the lever (a 1:5 ratio is convenient). The drum is stationary during the stimulation so that the contracting muscle writes only a vertical line. After each stimulation rotate the drum about 2 cm and wait ½ min before again stimulating. Use a 10-g weight.

2. Adjust the stimulator voltage so that the closing or opening of the key causes no contraction. By gradually increasing the strength, a faradic current will be reached that is just capable of evoking a barely visible contraction. Note whether this occurred during the break or make.

3. Continue to increase strength of current by adjustment of dial on inductorium in order to produce nicely graded contractions and obtain make and break contraction. When no further increase in the make or break contractions is possible, the experiment is complete.

4. Label tracing; indicate M (make) and B (break) and distance between primary and secondary coil (read from dial 1, 2, 3, etc.) Also label liminal stimulus and point of maximal stimulus.

5. When the tracing is completed, measure the magnified height in millimeters of the break (or make) contraction and record in table form. Calculate in millimeters the real height of contraction. Measure the length of the muscle (not tendon) and calculate the percentage of shortening during the contraction.

6. On coordinate paper plot a curve showing the relation of the strength of stimulus and the height of contraction. Use the distance between coils if the inductorium is used or voltage if read directly as the abscissae and the height of contraction in millimeters as ordinates. Use the magnified rather than real height.

DETERMINATION OF REAL HEIGHT OF CONTRACTION (see example below)

1. Measure distance between axis and point of attachment of muscle to muscle lever.

2. Measure distance between axis and end of writing pen.

Distance Between Coils (cm) (or Voltage)	*Height of Contraction (mm)*		*% Shortage*
	Magnified	*Real*	

3. Measure height of magnified contraction from graph.

Sample calculation

$$5 : 100 \text{ mm} :: x : 20$$
$$100x = 100$$
$$x = 1 \text{ mm real height}$$

Shortening

Measure length of muscle at rest.

Example

Muscle at rest = 10 mm in length
Muscle shortened 1 mm
% shortening = 10%

$$\% \text{ shortening} = \frac{\text{real height of contraction}}{\text{length of muscle at rest}} \times 100$$

$$\frac{1}{10} \times 100 = 10\%$$

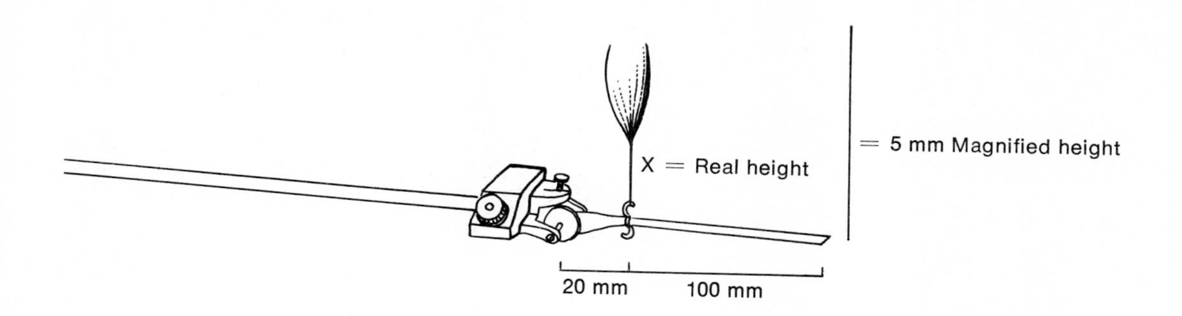

4. When using muscle contraction module note amplification position on bar and divide magnified height of contraction by amplitude used, that is, $10x$, $16x$, $24x$, or $32x$ to determine real height of contraction.

Experiment IV

Duration of Muscle Twitch and Influence of Temperature on Muscle Activity

APPARATUS

Stimulating and recording apparatus, tuning fork, thermometer, cracked ice, beakers, thread, Harvard muscle warmer, signal magnet, Ringer's solution.

PROCEDURE

1. Set up apparatus as in previous experiments. Find a strength of faradic current that evokes a good make contraction. Allow the drum to rotate at its fastest speed; if you find this too slow, twirl it by hand. Use fastest speed of chart mover. Keep the key closed in the primary circuit so as not to obtain a double twitch, and at the same time make sure the drum is stopped in time so that a doubling up of the curve is not obtained. After inscribing a good form curve, place the muscle pen in its original position and describe a base line. After developing the base line, turn the stand slightly away from the drum; rotate drum so that the muscle pen is just opposite the highest point of the form curve (top of contraction). Carefully apply muscle pen to this point of the tracing and allow it to fall gently to the base line. An arc will be described that will produce two points (one at the top of the curve and one at the bottom – base line), which are your simultaneous ordinates. It is to be noted that contraction does not occur immediately with the presentation of an adequate stimulus. A definite time is necessary for the chemical and physical processes that result in contraction to take place. This time is referred to as the **latent period**. It may be measured from your muscle graph by noting the distance between the signal magnet mark and the beginning of the muscle lever on its upswing. The distance between the start of the curve and the bottom of the arc is the duration of the **contraction period**, and the distance from the bottom of the arc to the point where the curve ends is the duration of the **relaxation period**. Remove your tracings, label all points, and calculate and record in 1/100 sec the length of contraction and relaxation. Also indicate the latent period. Record in table form.

	1/100 sec
Latent period	
Contraction period	
Relaxation period	

2. In the second portion of this experiment use the same muscle if possible. Use the same strength of stimulus and muscle load (5-g weight). Use muscle warmer for the attachment of muscle. To horizontal bar of muscle warmer attach femur of muscle preparation. The tendon is attached to a thread, which in turn is drawn through opening of the muscle warmer and attached to pulley of muscle lever so that upon contraction the muscle lever is lifted. Attach two fine wires, one to the femur end of the muscle preparation and the other to the tendon. These wires are to be connected with the wires from the secondary coil. Fill a beaker with ice, water, and some salt to lower freezing point; cool muscle to approximately 5° C by lifting beaker up and around muscle warmer. Liquid will not touch muscle so that Ringer's solution is not necessary. Obtain a muscle curve at 5° C. Repeat at 30° C. Record data in table form.

This will be a group tracing. Each student will individually discuss results.

	Duration of Contraction Components in 1/200 sec			*% Increase or Decrease from Room Temp.*
	5°C	*30°C*	*Room Temp*	
Latent period				
Contraction period				
Relaxation period				
Total contraction				

Experiment V | Muscle Work—Optimum Load

APPARATUS

Stimulating and recording apparatus, thread. Weights: 10, 30, 50, 70, 90, 100 g.

PROCEDURE

1. To stimulate muscle, use a strength of faradic current that will give a good break contraction. Always short circuit make stimulus. Use a constant strength of current. Make sure you afterload the muscle with the afterload adjustment screw. This will always bring the muscle lever into a horizontal position. The weight added is not supported by the muscle, and the muscle is not stretched until it begins to contract. The drum, stationary while the muscle is stimulated, is moved only between periods of stimulation. Record two contractions with each load to get an overall average contraction. Record a contraction with no load (neglecting weight of lever), then use 10-, 30-, 50-, 70-, 90-, and 100-g weights successively, getting individual contractions with each weight. If muscle still contracts, increase load by 50 g until no contraction is elicited by muscle. Disregard too feeble a contraction, that is, one that does not register more than 1 mm. Two types of muscular contraction are recognized. If the muscle lifts a load through a distance, work is accomplished and the contraction is called **isotonic**. If the load is too heavy to lift and the muscle is prevented from shortening when it contracts, the contraction is called **isometric**. During normal activity muscular contractions may be of both types, the loads being the bones of the body. Work is determined by multiplying the weight of the load lifted by the real height to which the load was lifted. Work is usually recorded in units of foot-pounds, kilogram-meters, or gram-millimeters. Work = real height of contraction times the weight of load.

   Example

   work = 2 mm × 50 g
   work = 100 g-mm

2. To study the effect of stretching, release the afterload so that the resting muscle is stretched. Bring muscle pen to horizontal position by raising muscle clamp. Stimulate

muscle and continue to add weights until no contraction is shown. Record these contractions on the kymograph. Indicate by appropriate labels what these contractions represent. It is to be recognized that a fundamental property of muscle fibers is that, within limits, increased tension (stretching) results in stronger contractions. Starling's law of the heart points out this factor, and it also applies to all muscles.

3. To determine maximum muscle change to maximal shocks, run your inductorium dial to ten units. The least load that the muscle you have been experimenting with is unable to lift expresses maximum strength.

4. The optimum load is the load at which the most work is performed by the muscle. Tabulate results in table form; that is, calculate work done moving various loads.

Indicate by underlining in red pencil the load that expresses maximum strength and the optimum load. Prepare a graph showing the relation of work done (ordinates) to load lifted (abscissae). On the same paper, plot a curve (in different color) of height of contraction against the load, using the same scale. Discuss the relationship between height of contraction and load and between work done and load.

Load (g)	*Height of Contraction (mm)*		*Work (g-mm)*
	Magnified	*Actual*	

Experiment VI | Effect of Load on the Components of a Muscle Contraction

APPARATUS

Stimulating and recording apparatus; weights.

PROCEDURE

1. Obtain a form curve as in previous experiments with a 10-g load. Obtain another form curve with the muscle loaded with a 50- or 100-g weight, according to size of muscle. Fill in the following table with data obtained:

	Duration in 1/100 sec		*Decrease or Increase with Change in Weight*	
	10 g	*50 or 100 g*	*in 1/100 sec*	*in %*
Latent period				
Contraction				
Relaxation				
Total twitch				

2. Discuss experiment in writeup, indicating which phase was most influenced by increase in load. Measure the height of magnified contractions. How were these influenced by load?

   With a single stimulus of adequate strength, increasing the load will result in stronger contractions up to a limit. There is a maximum pull or tension a muscle can exert, and loads greater than this result in weaker contractions.

Experiment VII | Fusion and Summation of Twitches

APPARATUS

Stimulating and recording apparatus, thread, 5- and 10-g weights.

PROCEDURE

1. If, instead of single shocks, repetitive stimuli are applied to a muscle, a series of twitches will be produced. If the repetition is rapid enough, the muscle will not relax between stimuli and the twitches will fuse with one another. The fused twitch will usually be higher than the preceding one, even though the strength of the stimuli is the same. As in previous experiments, obtain muscle curves, using in this experiment a faradic shock that gives a maximal make and break contraction. Use a 5- or 10-g weight on muscle and medium speed drum. After several individual muscle contractions are recorded, progressively shorten the time between the make and break stimulus until, in the last curve, the second twitch begins before the relaxation of the first twitch has set in. The finished graph should have a series of muscle twitches showing the progressive shortening of time between stimuli (muscle curves closer together) until summation is actually reached.

2. Label tracing "Summation of Twitches."

3. Discuss your results. Compare heights of single twitch with height of contraction when summation took place. How do we account for summation? Is the greater height in the second twitch caused by a greater number of muscle fibers thrown into action?

Experiment VIII | Incomplete and Complete Tetanus

APPARATUS

The usual stimulating and recording apparatus. Use the vibrating interrupter on inductorium in this experiment. Thread and 5- and 10-g weights.

PROCEDURE

1. Use a strength of current that gives make and break twitches of the same height. It is important not to fatigue the muscle. The speed of the drum should be fairly rapid and a 5- or 10-g weight attached to the muscle. Send a series of ten shocks into the muscle at the rate of one per second. When the lever has reached the base line, stop the drum and let muscle rest for at least 1 min. Start the drum and stimulate again with 20 shocks at 1/2-sec. intervals. Increase rate until maximum number of stimuli capable of being produced by this method is reached. Label graph "incomplete tetanus." This is a group tracing.

2. Now slow the drum somewhat. Stimulate the muscle with a very rapid series of shocks given by the inductorium wired for multiple stimulation. This amounts to about 60 stimuli per second. Hold the stimulation for about 10 sec and let the drum run until the pen reaches the base line. Label tracing "tetanus." Each member of the group will make a tracing of tetanus.

3. Discuss the data from the experiment in your writeup. In what respect does incomplete differ from complete tetanus? How can we account for the smooth curve in complete tetanus? Does the law of all or none apply?

Experiment IX | Staircase and Fatigue

In the early stages of a series of muscle twitches an interesting phenomenon called **staircase**, or **treppe**, can be observed. If a series of stimuli of sufficient strength is sent into a muscle, so that each contraction can be fully recorded, that is, about one per second, the succeeding contractions, after the first, increase gradually in height even though the stimulus is of the same strength. This makes the myographic record appear like a staircase. The contractions will level off at a plateau, and upon continual stimulation each succeeding contraction will become weaker until finally no response will be elicited, no matter how strong the stimulus applied to the muscle. This, referred to as **fatigue**, represents a loss of irritability by the muscle.

APPARATUS

Stimulating and recording apparatus, weights, litmus paper – red and blue

PROCEDURE

1. Use a faradic current strong enough to give a good break but not make contraction. Use a 6 or 10 g load, making sure you afterload the muscle. Use a slowly moving drum and stimulate the muscle about 15 times at the rate of two stimuli per second. Stop the drum. Increase the strength of stimulus considerably and immediately repeat stimulation until fatigue is again experienced. Now remove the load and again cause fatigue. This will be a group tracing.

2. Label tracing showing the following effects:

   a. The beneficial effect of previous work (staircase)
   b. The gradual decrease in the power of the muscle as it continues to work (fatigue)
   c. That fatigue is relative to the strength of the stimulus and the amount of load moved

3. Remove the fatigued muscle and cut it into two parts. Press the cut surface of one piece on blue litmus and the other cut surface on red litmus. Try this with a fresh muscle. Explain the results. Source of material responsible for this reaction? How is this material formed? What normally becomes of it? Why didn't this happen in the experiment? Why is it said that fatigue is autointoxication?

Experiment X | Influence of Fatigue on Muscle Activity

In this experiment the form curve of a fatigued muscle is to be compared with that of a rested muscle.

APPARATUS

Stimulating and recording apparatus, weights, tuning fork, signal magnet

PROCEDURE

1. Set up muscle preparation as usual. Stimulate the muscle with a strong break faradic shock, short circuiting the makes. Use a 20- to 80-g weight according to the size of the muscle. Have the drum high on its axle and the pens as low as possible. Place the tuning fork pen directly below the muscle pen and obtain a form curve as you did in a previous experiment. Partially fatigue muscle by stimulating it with a tetanic current for 2 sec. These are not to be recorded. Having lowered the drum to clear the first tracing again, obtain a form curve. Repeat this procedure of fatiguing muscle and then securing form curves until you have two or three more curves. The last curve should allow a marked degree of fatigue. This will be a group tracing. Label "Influence of Fatigue." Discuss your results.

2. Tabulate your results.

	Duration in 1/100 sec		
Muscle Curve	*Rested*	*Fatigued*	*% Change*
Latent period			
Contraction			
Relaxation			
Total twitch			

28

Integument

If we examine the skin by histological methods, cutting out a thin vertical slice and staining it, we discover that it consists of two main divisions. The innermost layer, called the **dermis** (Gr., skin), is made of fairly loose connective tissue and contains many blood vessels and nerves. The outermost division, or **epidermis** (Gr. upon skin), is devoid of blood vessels and has nerves only in its deepest parts. It is continually formed by the growth of its deepest so called germinative layer. The cells are continually pushed outward through several zones, and finally become layers of horny scales that protect us from our external environment.

The structure of the epidermis is most typical where it is thickest. In sections perpendicular to the surface four main layers can be distinguished (Fig. 44).

1. The deepest layer, which touches the derma, is known as the **stratum germinativum**; it consists of cylindrical cells, which become flattened out as they go toward the granular layer. The surface of these cells is covered with thin spines.
2. The granular layer (**stratum granulosum**) consists of three to five layers of flattened cells, which are rhomboidal in a perpendicular section.
3. The clear layer (**stratum lucidum**) consists of several layers of flattened, closely packed cells, which in section appear as a pale wavy stripe.
4. The horny layer, the **stratum corneum** on the palms and soles, reaches a considerable thickness and consists of dead, cornified, flattened cells. This is the outermost layer.

On most of the body the epidermis remains much thinner and has a much simpler structure: two layers are always present, the germinating layer and the stratum corneum.

In the dermal layer we find there is a wide variety of bulbs, corpuscles, and naked nerve endings (special staining methods must be used to bring these out). In addition, the hair roots, set down in pits or follicles, can be seen. Oil glands on one side of the hair root can be recognized – the **sebaceous glands.** On the same side is a smooth muscle band, the arrector pili, the main function of which is to erect the hair (Fig. 45).

Blood vessels also can be seen.

Draw a section of human skin from the slides provided. Indicate all parts that can be seen.

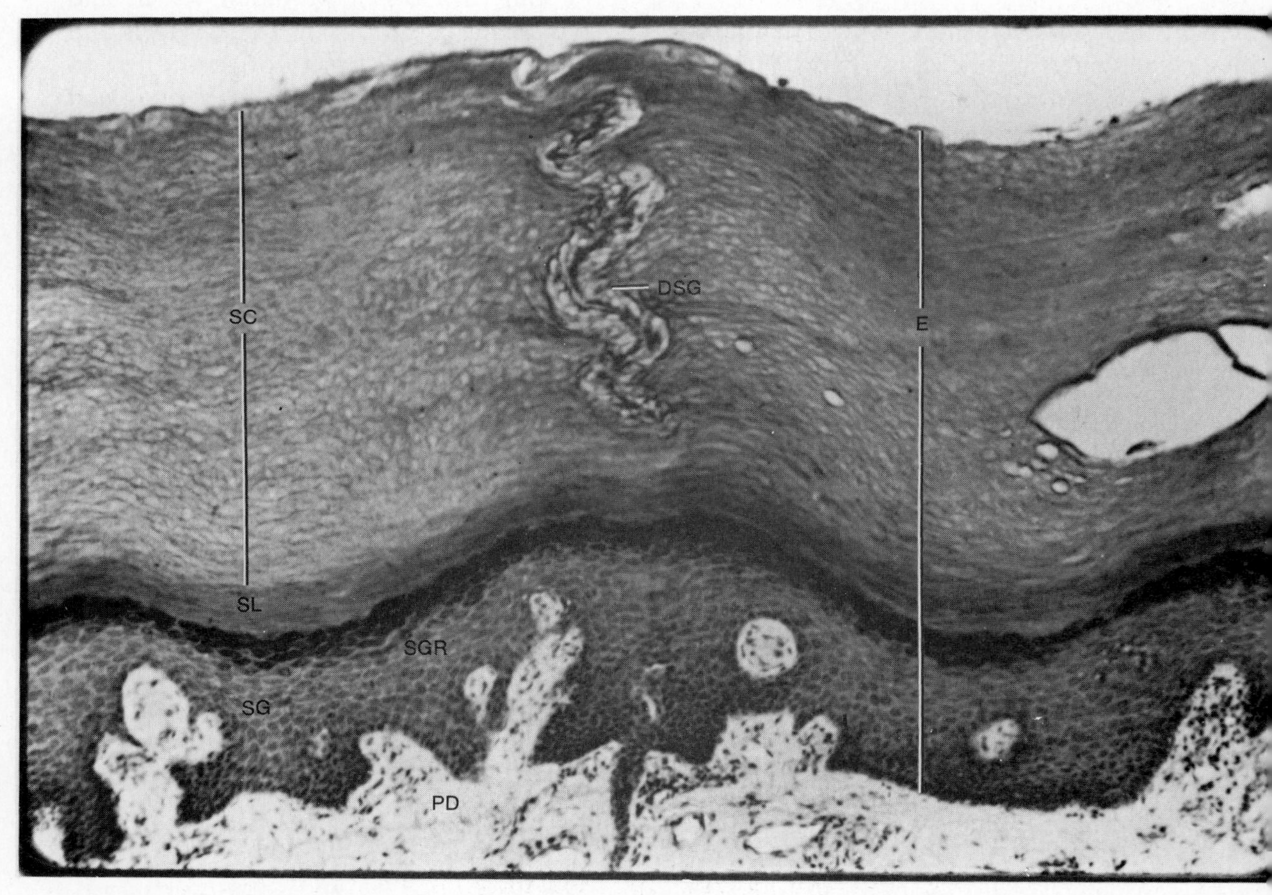

Fig. 44. Cross section of cornified skin from sole of foot showing structure of epidermis. E, epidermis; DSG, duct of sweatgland; SC, stratum corneum; SL, stratum lucidum; SGR, stratum granulosum; SG, stratum germinativum; PD, papillary layer of dermis.

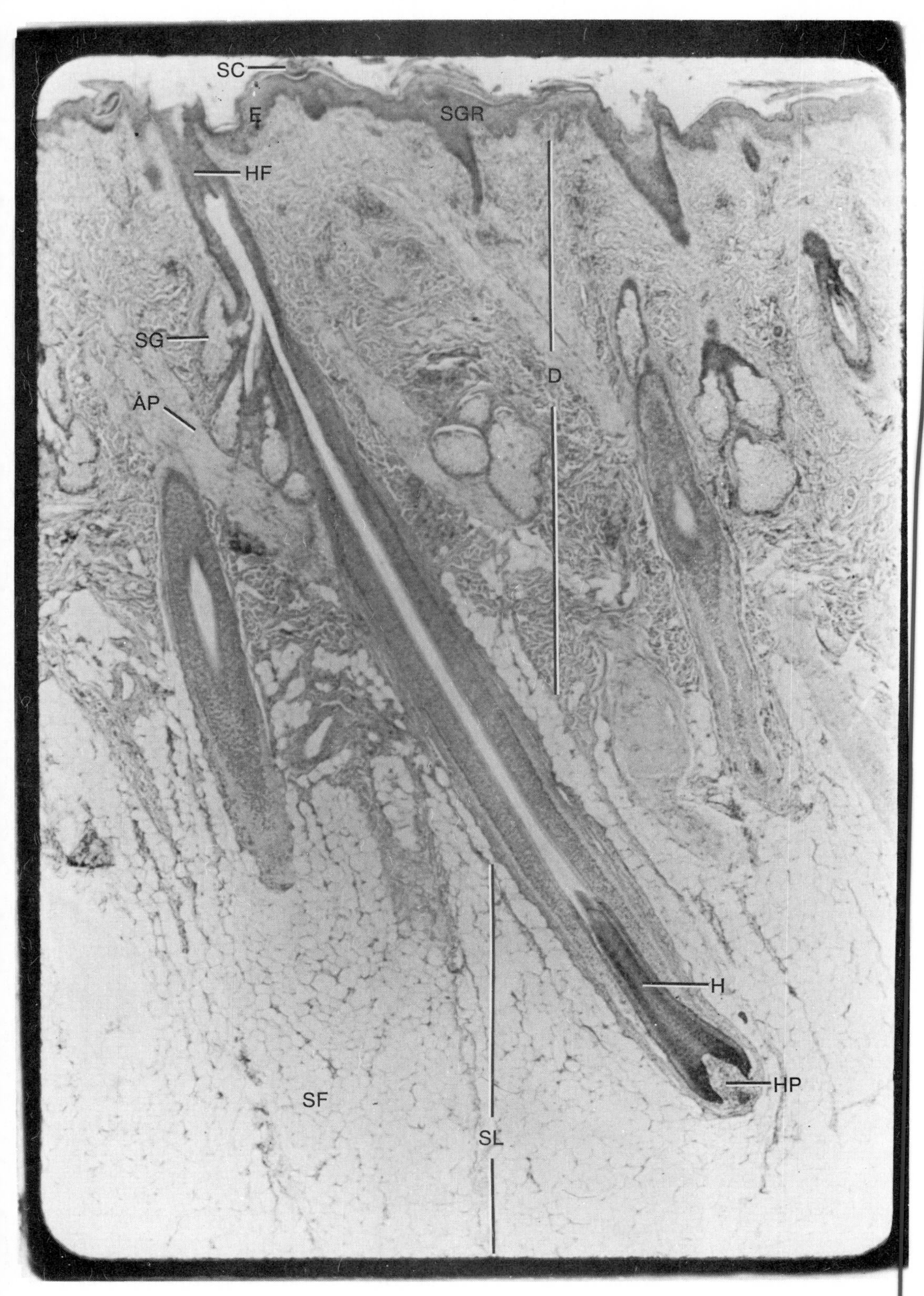

Fig. 45. Skin of scalp showing detail of dermal layer. SC, stratum corneum; E, epidermis; SGR, stratum granulosum; HF, hair follicle; SG, sebaceous gland; AP, arrector pili — hair muscle; D, derma, H, hair; HP, hair papilla; SL, subcutaneous layer; SF, subcutaneous fat.

Label the following parts:

1. Epidermis
2. Various layers of epidermis that can be seen
3. Dermis
4. Smooth muscle
5. Sebaceous gland
6. Hair papilla
7. Hair shaft
8. Blood vessels
9. Fat cells in deep dermal layer

29

Brain

Dorsal Aspect of the Brain

The skull must now be opened and the brain exposed (Fig. 46). First remove all muscles and the external ears. Then make an incision (with the bone saw) along the lateral line of the dorsal part of the skull on each side; make another across the skull just back of the nasal bones; make a fourth incision across the skull just above the occipital condyles. Be very careful not to saw into the brain; saw only a short time in one place, then move the saw a bit farther, being careful to follow the curvature of the skull. Then take a blunt instrument such as the handle of a scalpel or a screwdriver, and pry out the roof of the skull. The bones from one side of the skull should next be removed with bone forceps, but be careful not to injure the cranial nerves. Be sure to keep the bones of the ear, and the eyes for later study of sense organs. If possible, it is better to use preserved specimens of sheep brains for this exercise. See Fig. 47 and 48.

1. **Meninges** – the membranes covering the brain and spinal cord:
   a. **Dura mater** – tough lining inside the skull. It grows downward between the hemispheres forming the falx cerebri. It is ossified between the cerebrum and cerebellum; this part is called the **tentorium cerebelli.**
   b. **Subdural space** – the space inside the dura mater.
   c. **Arachnoid** – a very delicate membrane in the subdural space. It does not enter the convolutions of the brain.
   d. **Pia mater** – a thin membrane covering the brain and inside the subdural space. The subdural space is between the dura mater and the pia mater. The pia mater enters all the convolutions of the brain.
2. **Olfactory bulb** – at the anterior end of the brain.
3. **Cerebrum** – composed of two hemispheres.
4. **Convolutions** – the grooves are called **sulci** and the ridges **gyri** (singular, sulcus and gyrus).
5. **Cerebral fissure** – the deep groove between the cerebral hemispheres. Also referred to as great longitudinal fissure.
6. **Corpus callosum** – a band of nerve fibers between the cerebral hemispheres; this band connects them. Spread the hemispheres apart to see these fibers.
7. **Cerebellum** – posterior to the cerebrum and composed of two hemispheres and a central part, or vermis.

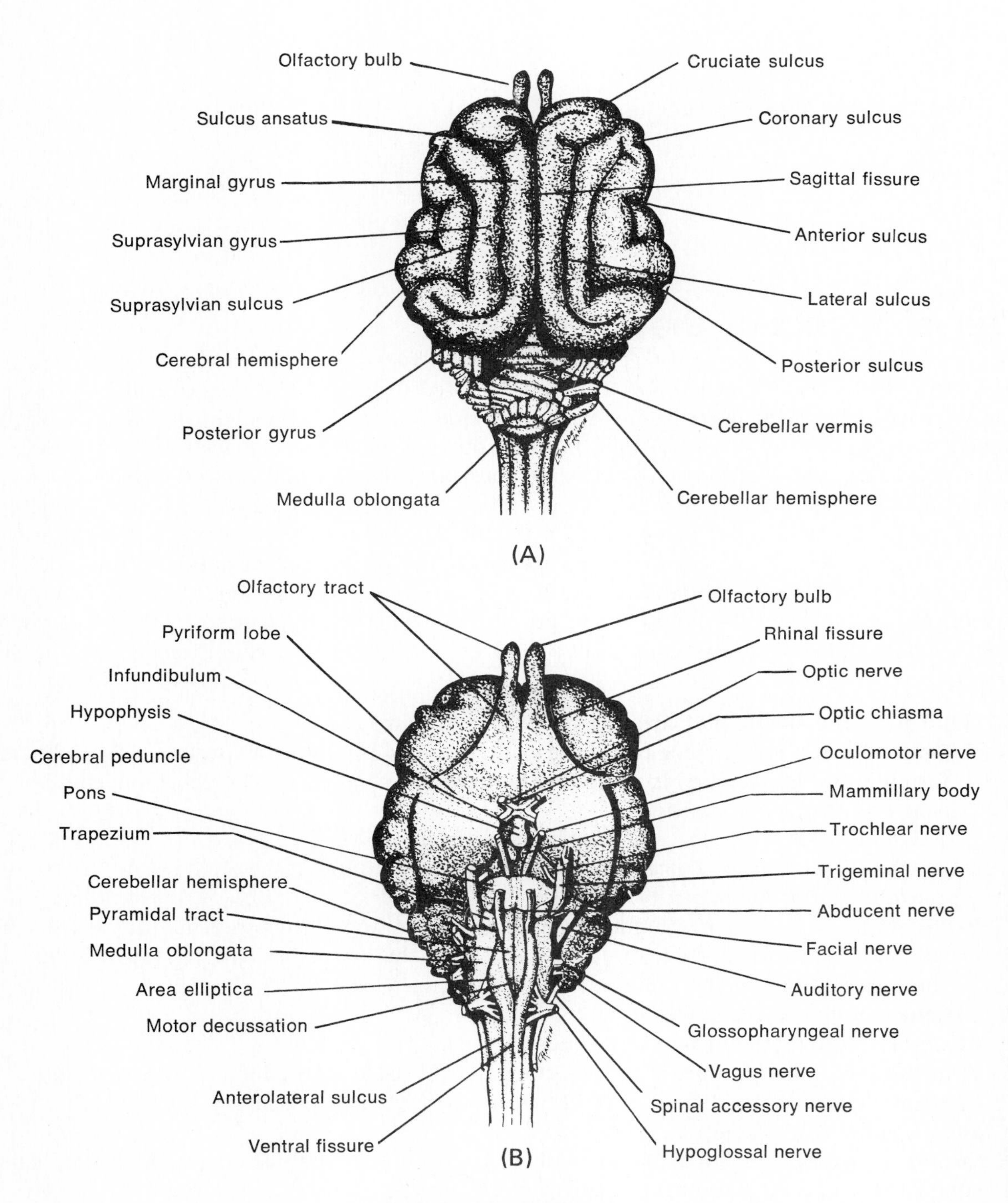

Fig. 46. Brain of the cat.
A. Dorsal aspect of the brain.
B. Ventral view.

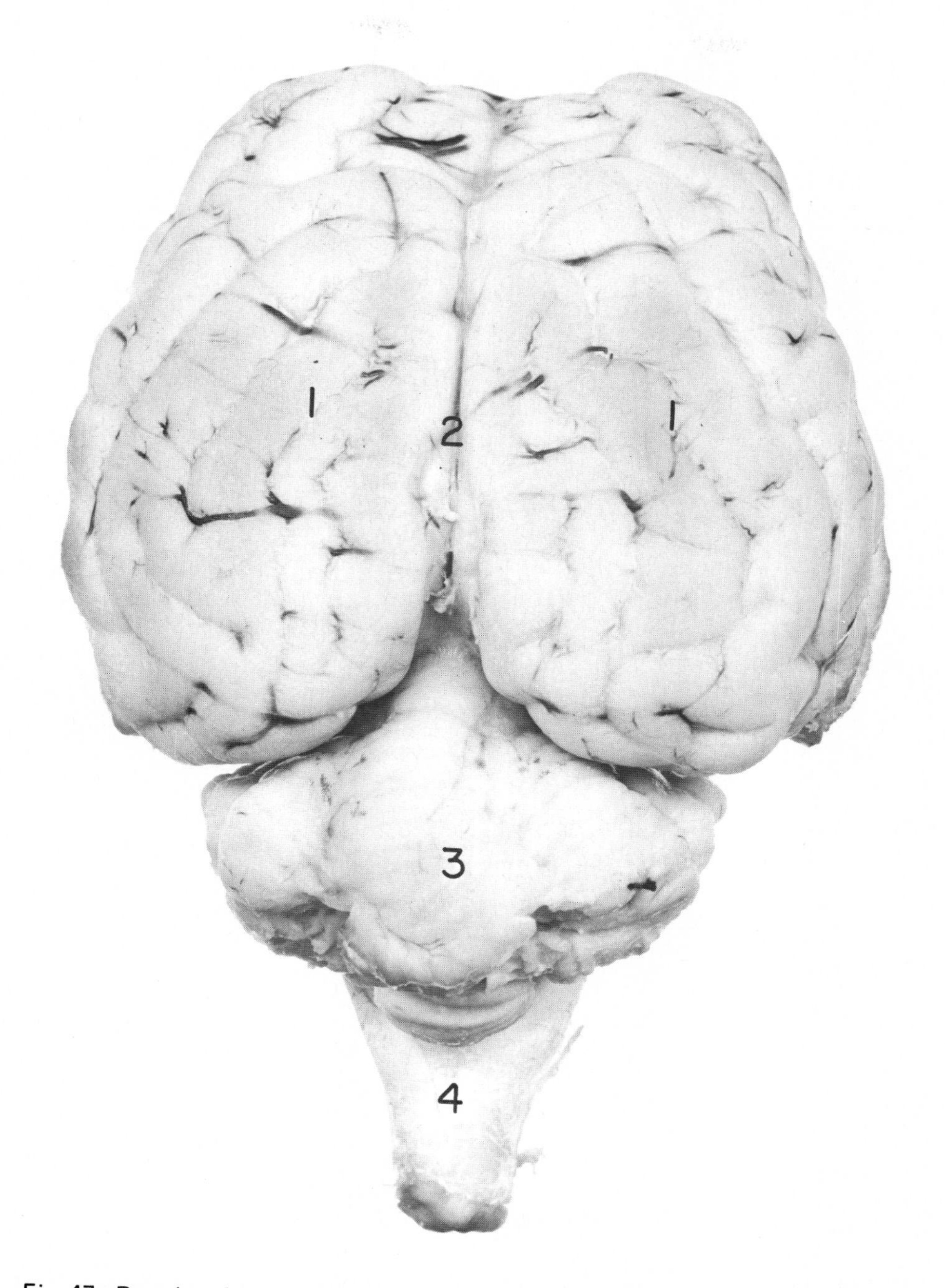

Fig. 47. Dorsal surface of sheep's brain. The cerebral hemispheres and cerebellum make up practically the entire aspect of the brain from this view.

1. Paired cerebral hemispheres.
2. Longitudinal cerebral fissure.
3. Cerebellum.
4. Medulla oblongata.

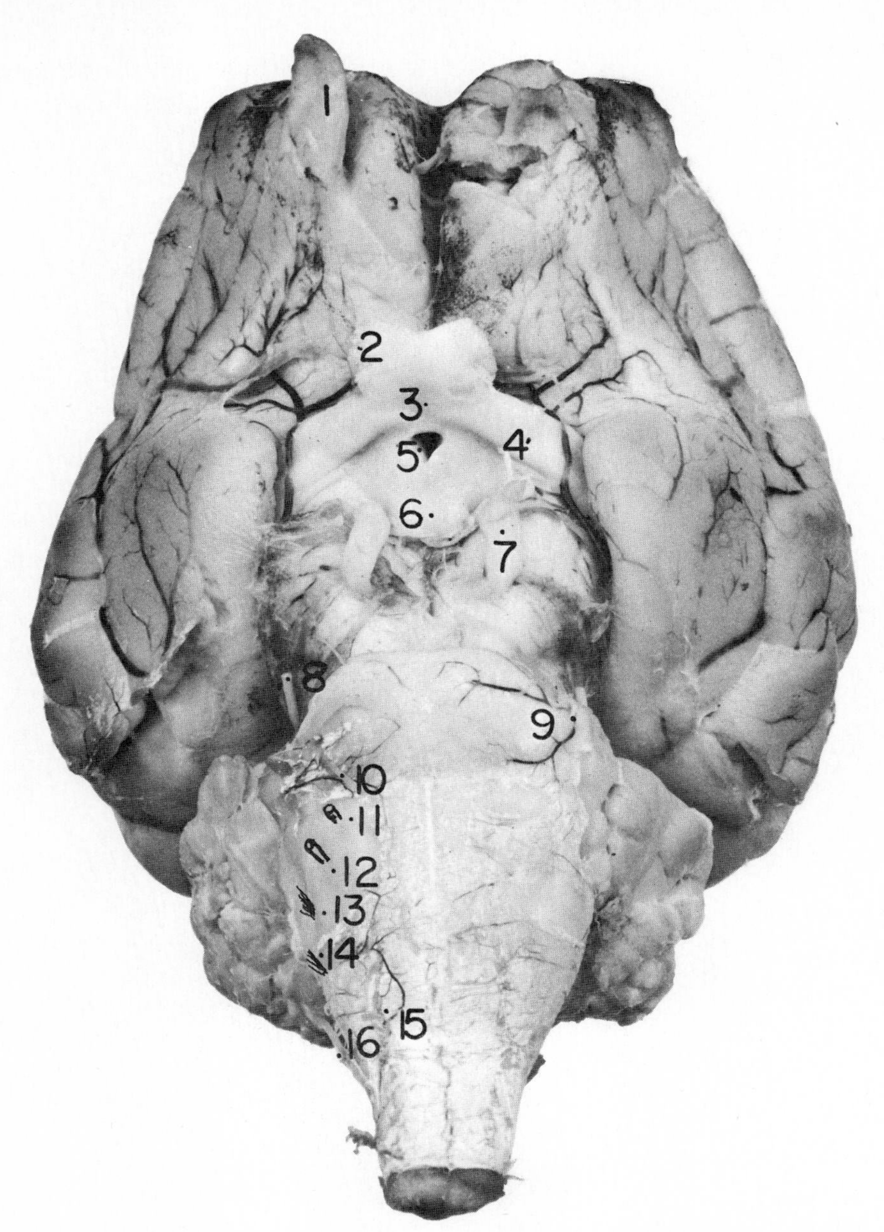

Fig. 48. Ventral surface of sheep's brain. The pituitary gland may still be suspended by a narrow stalk (infundibulum) from the hypothalamic area. Remove this structure and note the cavity in the infundibulum, which represents an extension of the third ventricle.

1. Olfactory bulb	I	9. Trigeminal nerve	V	(stump is seen)
2. Optic nerve	II	10. Abducens nerve	VI	
3. Optic chiasma		11. Facial nerve	VII	(drawn in)
4. Optic tract		12. Acoustic nerve	VIII	(drawn in)
5. Infundibulum		13. Glossopharyngeal	IX	(drawn in)
6. Mammillary body		14. Vagus	X	(drawn in)
7. Oculomotor nerve	III	15. Spinal accessory	XI	
8. Trochlear nerve	IV	16. Hypoglossal	XII	

Note: On commercial specimens, all the nerves are not demonstrated. The photograph has been altered to show the general idea and location of the cranial nerves.

8. **Midbrain (mesencephalon)** – located between the cerebellum and the cerebrum. It is made up of four lobes called **corpora quadrigemina,** or **colliculi.** The anterior two lobes are superior colliculi, and the posterior two are inferior colliculi. The midportion of the mesencephalon is known as the **tegmentum** and contains important motor and sensory pathways.
9. **Medulla (medulla oblongata** or **myelencephalon)** – part covered by the cerebellum. The lower part of the skull and the anterior part of the vertebral column will be removed to make this plain.
10. **Fourth ventricle** – a cavity in the medulla just below the cerebellum. It is wider than but continuous with the cavity of the spinal cord.

Ventral Aspect of the Brain

Be sure that all the bones on one side of the skull are removed and that the cranial nerves are intact. Dissect out the cranial nerves on one side, loosen the anterior part of the brain carefully, and then gradually work back until the whole brain can be lifted out. Be sure that the optic chiasma and hypophysis remain attached to the brain. They are located about one third of the way back. Cut the spinal cord posterior to the medulla and remove the brain into a receptacle of water (do not have the water deep enough to cover the brain). Find the following structures:

1. **Olfactory bulbs** – already mentioned above.
2. **Olfactory tracts** (first) – bands of fibers extending back from the bulbs and ending in the pyriform lobe near the optic chiasma.
3. **Optic chiasma** – the area where the optic nerves (second) come together with part of their fibers crossing from one side to the other.
4. **Tuber cinereum** – a small round knob just posterior to the optic chiasma.
5. **Infundibulum** – extends from the center of the tuber cinereum.
6. **Hypophysis** or **pituitary body** – on the posterior and ventral side of the infundibulum.
7. **Mammillary bodies** – fastened to the posterior part of the tuber cinereum and partly covered by the infundibulum.
8. **Posterior perforated space** – depression around the tuber cinereum.
9. **Cerebral peduncles** – a bundle of fibers located in the lateral part of the posterior perforated space.
10. **Oculomotor nerve** (third) – arises from the side of the cerebral peduncle.
11. **Trochlear nerve** (fourth) – arises from the dorsal side of the midbrain and appears from between the cerebellum and the inferior colliculi.
12. **Pons varolii** – a heavy transverse band of nerve fibers posterior to the posterior perforated space.
13. **Brachium pontis** – a more slender continuation of the pons varolii extending dorsally to the cerebellum.
14. **Trigeminal nerve** (fifth) – arises from the lateral part of the ventral surface of the medulla just posterior to the pons.
15. **Ventral fissure** – groove in the center of the medulla.
16. **Pyramidal tracts,** or **pyramids** – a bundle of fibers on each side of the ventral fissure, becoming narrower toward the posterior end.
17. **Trapezoid body** – a narrow transverse band of nerve fibers passing under the pyramids.

18. **Abducens nerve** (sixth) – arises at the edge of the trapezoid body where it disappears under the pyramids.
19. **Facial nerve** (seventh) – arises next to the root of the trigeminal nerve posterior to the pons.
20. **Auditory nerve** (eighth) – arises posterior to the root of the trigeminal nerve.
21. **Glossopharyngeal nerve** (ninth) – arises just posterior to the auditory nerve.
22. **Vagus nerve** (tenth) – arises just posterior to the glossopharyngeal nerve.
23. **Spinal accessory nerve** (eleventh) – arises just posterior to the vagus nerve.
24. **Hypoglossal nerve** (twelfth) – arises posterior to the spinal accessory and is at the posterior limit of the medulla.

30

Nerve Histology

Nerve cells carry nerve impulses to all organs of the body. A prepared slide of a cross section of the nerve cord of the cat is to be used for the study of nerve cells. Near the center of such a section cut through the nerve cord and look for large cell bodies showing **nerve processes** radiating from them. Note model of a nerve cell. The nucleus can be seen in the center of the cell body. A single nerve cell is called a **neuron.** The cell membrane is called **neurilemma** and the cytoplasm **neuriplasm.** Generally, one **axon** carries impulses away from the cell body, and many **dendrites** carry impulses toward the cell. A **sensory neuron** receives stimuli and transmits them toward the spinal cord or brain, whereas a **motor neuron** carries impulses from the brain or spinal cord to muscles, causing them to contract, or to glands, causing them to secrete.

Draw a nerve cell showing the processes and nucleus.

On either side of the nerve cord, on the same slide, are masses of nerve cells whose processes do not show here. They are cross sections of the cells in the dorsal root ganglion, leading to the dorsal part of the spinal cord. In this region look for the largest roundish cells that can be found. These are sections of the nerve cell centers. Each cell has a large nucleus showing clearly the chromatin granules, nucleolus, and the nuclear membrane. The **nucleolus** (little nucleus) is the large, deeply stained granule in the nucleus. Does the cytoplasm appear to be shrunken away from the limiting boundary? Note the row of nuclei around each cell. These are the nuclei of small **satellite cells,** which serve to protect the nerve cell. Can you see the cell membranes of the satellite cells? (Refer to your text, *The Human Body: Its Structure and Function,* for detail of structure.)

Draw one of these nerve cells showing its parts and the surrounding satellite cells. Make the drawing about 1½ in. in diameter.

31

Spinal Cord

Remove the skin and muscles from a part of the spinal column; then with bone forceps cut off the neural spines and arches of two vertebrae, thus exposing the spinal cord in place. Locate all the structures called for below; then remove a piece of the cord and study its cross section. You will notice the H shaped appearance of the gray matter in contrast to the white matter. Locate the following:

1. **Spinal nerves** – for each nerve, the dorsal root emerges from the dorsal part of the spinal cord, and the ventral root emerges from the ventral part.
2. **Spinal ganglion** – a swelling on the dorsal root of the spinal nerve.
3. **Dorsal median sulcus** – the dorsal groove in the center of the spinal cord.
4. **Dorsolateral sulcus** – a groove on either side of the dorsal median sulcus.
5. **Ventral fissure** – the groove in the midventral part of the spinal cord.
6. **Dorsal funiculus** – the tissue between the dorsal median and dorsolateral sulci.
7. **Lateral funiculus** – tissue between the dorsolateral sulcus and the point where ventral root nerves emerge.
8. **Ventral funiculus** – tissue between the ventral root and the ventral fissue.

Spinal Nerves

Spinal nerves are those emerging from one part or another of the spinal cord. In the cat there are 38 or 39 pairs of them: 8 cervical, 13 thoracic, 7 lumbar, 3 sacral, 7 or 8 caudal. They emerge from the spinal cord through the intervertebral foramina. Locate the following:

1. **Cervical spinal nerves 1 to 4** – these serve the muscles of the lateral anterior part of the neck.
2. **Cervical nerves 5 to 8** – these send branches into the brachial plexus. From the ventral rami of nerves 5 and 6 come the branches that make up the phrenic nerve (extending to the diaphragm). Trace this nerve from the spinal cord to the diaphragm.
3. **Brachial plexus** – a network of nerves that arises from the fifth, sixth, seventh, and eighth cervical nerves and the first thoracic nerve. Remove the skin, cut the pectoral muscles, and separate them from the serratus muscles; then remove the pectoral muscles from the chest

wall and the upper arm. The brachial plexus and all the following nerves should be exposed. Locate them and trace their origins in the spinal nerves mentioned above and their endings in the arm: fifth, sixth, seventh, and eighth cervicals; first thoracic; phrenic; suprascapular; musculocutaneous; ventral thoracic; axillary; radial; median; ulnar; medial cutaneous; long thoracic or dorsal thoracic; first, second, and third subscapulars.

4. **Thoracic nerves 2 to 13.**
5. **Lumbar nerves 1 to 7.**
6. **Lumbosacral plexus** – formed by the ventral rami of the fourth, fifth, sixth, and seventh lumbar nerves and the three sacral nerves. The plexus is located between the iliopsoas and the psoas minor muscles. Lift the visceral organs to one side, but do not remove them. Remove all the iliopsoas and psoas minor muscles necessary to expose the plexus. Locate the following nerves: first, second, and third sacral nerves; sciatic; genitofemoral; lateral femoral cutaneous; femoral; long saphenous; obturator; gluteals; common peroneal; tibial; posterior femoral cutaneous; pudendal; and the inferior hemorrhoidal nerves.
7. **Sacral nerves 1 to 3.**
8. **Caudal nerves 1 to 8** – locate one or two of these.

Humans have 31 pairs of spinal nerves named for the vertebrae anterior to their emergence. There are 8 cervical, 12 thoracic, 5 lumbar, 5 sacral, and 1 coccygeal.

32

Sensory Input of Spinal Nerves — Dermal Segmentation

Sensory input from the environment is carried by the spinal and some cranial nerves to the brain. It is through the dorsal root of the spinal nerve that sensory impulses enter into the central nervous system. In early embryonic life, while the spinal cord still occupies the entire length of the vertebral canal, the spinal nerves pass directly out to the dermatomes of that level. As the extremities develop and the body takes on its adult form, there is a migration of the dermatomes to form the subcutaneous connective tissue and the dermis of the skin. The differentiating dermatomes, however, carry with them their earlier established innervation, so there is a definite segmental pattern of the cutaneous distribution of various nerve roots, one segment lying behind the other in an orderly sequence. Some knowledge of the dermatome pattern is important as it is possible to localize the size of lesions involving the spinal cord. Charts outlining the exact sensory dermal segments are given in most neuro-anatomy and neurophysiology texts. Some surface landmarks on the body that will serve as a general guide to localization are listed in Table 1.

Table 1. *Dermatome Innervation Patterns*

Cutaneous Region of Body Supplied	*Sensory Root of Spinal Nerve*
Trunk region at level of clavicle	5th cervical
Trunk region at level of nipple	4th thoracic
Trunk region at level of umbilicus	10th thoracic
Groin region	12th thoracic
Deltoid region	5th and 6th cervical
Lateral aspect of arm	5th, 6th, and 7th cervical
Inner aspect of arm	8th cervical–1st thoracic
Hand from radial to ulnar borders	6th, 7th, 8th cervical
Outer and dorsal surface of thigh	5th lumbar and 1st, 2nd sacral
Inner and ventral surface of thigh	1st, 2nd, 3rd, 4th lumbar
The perineum	2nd, 3rd, 4th, and 5th sacral
The foot from lateral to medial surface	1st sacral, 5th, and 4th lumbar

Experiment XI | Function of Nerve Fibers

APPARATUS

Stimulating apparatus, glass rod, NaCl crystals, 1 percent HCl solution, bunsen burner, ring stand, beaker of water, femur clamp, frog board.

In handling nerves never pinch or stretch the nerve or handle it with metals. Keep it moist with Ringer's solution.

PROCEDURE FOR MAKING A NERVE MUSCLE PREPARATION

1. Pith a frog; hold frog in left hand after washing so that it is not too slippery. With the index finger press down well on the nare. Glide the nail of your right index finger over the bent head of the animal. A small groove will be felt about 3 mm back of the posterior borders of the eardrums. This marks the junction of the first cervical and the skull. Press a dissecting needle into the groove so that it passes forward into the foramen magnum into the skull. The brain can be destroyed by twisting the needle from side to side.

   If there is no reaction when the toes are pinched at this point, spinal shock is being evidenced. It will pass off within 5 or 6 min. The destruction of the brain can be determined by the corneal reflex. If corneal reflex is present (closing of lower lid when eye is touched), the brain is not thoroughly destroyed. When the animal is in spinal shock, you cannot tell; so after about 6 min test the toes by pinching and test corneal reflex. The former should elicit a response, the latter no response. By reversing the direction of the pin destroy the spinal cord.

2. After properly pithing the frog, cut away all the body in front of the urostyle with a large pair of scissors. Remove skin from the trunk and legs; slit open abdomen and remove organs. This will expose three roots of the sciatic nerve on each side of the vertebral column. Slip a moist thread under all three nerves fairly close to the vertebral column. Tie the ligature and cut between the ligature and the backbone. Use the ligature to lift nerve. Turn the frog over and with forceps grasp the lateral and medial muscles of the thigh and gently pull them apart. The sciatic nerve will show itself as a white glistening thread. Very carefully dissect away the muscle from the nerve and free it up to the knee. Now

dissect out the gastrocnemius muscle as you have been doing. The result is the nerve muscle preparation.

3. Keep the femur in a clamp and support nerve on a horizontally fixed glass slide. Keep it moist; no record is necessary.

   By noting contractions of the muscle, determine whether the following are stimuli.

   a. Mechanical–pinch free end of nerve. Result?
   b. Thermal–apply a hot glass rod to the end of the nerve. Result?
   c. Osmotic–apply a few crystals of NaCl to the end.
   d. Chemical–dip the end of nerve in the 1 per cent HCl solution. Result?
   e. Electrical–apply mild faradic shocks to the nerve. Discuss results and experiment.

   What proof of irritability is there? Of nerve conductivity? Of the efficiency of the various stimuli? Why would you use electric current in most experiments as a stimulus? How many of these stimuli is our body subjected to? What are a direct stimulus and an indirect stimulus? What is a nerve? What are the two outstanding physiological properties of a nerve fiber?

Experiment XII

Nerve Blocking

APPARATUS

Stimulating apparatus, glass plate, absorbent cotton, ether (or novocaine solution), filter paper. Platinum electrodes, frog board, dropper.

PROCEDURE

1. Make a muscle nerve preparation with the nerve as long as possible. Place the nerve on the points of the platinum electrodes and determine the liminal stimulus.

2. Lay the nerve preparation on a moist glass plate and place under the middle portion of the nerve a small wad of cotton soaked with ether or novocaine. Cover with another piece of cotton.

3. Place over the whole preparation – but not in contact with the nerve – a piece of moist (Ringer's solution) filter paper.

4. At 2-min intervals determine the minimal stimulus for the nerve by stimulating the free end of the nerve that has not been affected by ether. It may be necessary to moisten cotton with a few drops of ether.

5. When complete anesthesia has been produced, stimulate nerve between anesthetized area and muscle. Now wash away the ether with Ringer's solution and at 2-min intervals again determine the minimal stimulus. Tabulate your results.

Discuss your results. What is conductivity? What is meant by the minimal stimulus? Of what is it a measure? Of what use is nerve blocking? Cite an example.

Anesthesia		*Recovery*	
Min	*Threshold*	*Min*	*Threshold*
0		0	
2		2	
4		4	
6		6	
8		8	

Experiment XIII

Effect of Curare on Muscle Nerve

APPARATUS

Stimulating apparatus, 1 percent curare solution, thread, hypodermic syringe, 1 percent eserine salicylate.

PROCEDURE

1. Destroy the brain of a frog and fasten frog dorsal surface to a frog board.

2. Make a small longitudinal cut in the skin of the left thigh and break apart the thigh muscles to expose the sciatic nerve. Do not injure the blood vessels. Injury to blood vessels will ruin the experiment.

3. Isolate nerve for a short distance, place a thread under the nerve, and tie it tightly around all other structures in the thigh. This prevents circulation to the gastrocnemius muscle.

4. Using a hypodermic syringe, inject ½ ml of 1 percent curare solution into the dorsal lymph sac. Pick up skin in general area laterally to the urostyle and inject needle below surface of skin (not into the muscle). Wait 30 min for the curare to take effect. Loosen the legs and pinch the right; if a contraction occurs, the drug has not yet taken effect.

5. After the frog is well under the influence of curare, expose the right sciatic nerve. Place a thread under the nerve and tie the free ends. You now have a loop by which you can lift the nerve. Lift the nerve onto electrodes without the electrodes touching any other structure. Determine the liminal stimulus (induced current) for the right sciatic nerve. If the experiment is successful, no contraction of the muscle will occur.

6. Remove skin from both calf muscles. Stimulate muscle. Results?

7. Now stimulate the left sciatic nerve. Results?

8. With the frog still under influence of curare, inject into dorsal lymph sac 1 ml of 1 percent eserine salicylate. After a few minutes stimulate the right sciatic nerve. Result?

Discuss your results. What is your conclusion as to the action of curare on muscle tissue? On stimulating the left sciatic nerve? Why does this result differ from that obtained with the right nerve? What is eserine with respect to curare? What effect has curare on muscle? On nerve? Where must the seat of action of curare be?

Experiment XIV

Motor Areas of the Mammalian Brain (Optional Experiment)

The gray matter of the cerebral cortex is divided into motor, sensory, and association areas.* These areas have been studied by direct stimulation of the cortex by electrical shocks or by locally applied drugs (strychnine), which lead to local but purposeless movements of the involved portion of the body. Such studies, referred to as "mapping out of the brain," have led to the numbering of the cortex in areas 1 to 52. The motor cortex has been divided into subdivisions, of which areas 4, 6, and 8 have been most extensively investigated.

APPARATUS

Stands, clamps, stimulating apparatus, scalpel, ether, cotton toweling, dental drill or trephine, mature albino rats.

PROCEDURE

1. Place rat in ether jar (glass cylinder in which ether is poured onto a cotton swab in bottom) until deeply anesthetized. The student acting as anesthetist should be very cautious and watchful during this period of anesthesia so as not to overanesthetize and kill the experimental animal.

After induction, anesthesia may be maintained by means of a small glass funnel with ether soaked cotton placed over the animal's nose. The depth of anesthesia is controlled by moving the container on or off the animal's face. The coloration of the skin, especially of the paws, is a good indicator of depth of anesthesia. When the color becomes bluish, stop the anesthesia and reapply when a good pinkish color appears.

2. The operator, with a holding assistant to stabilize the animal's head, makes a midline incision in the scalp from a point just posterior to the nasal bones backward to the occiput. Bleeding in the scalp can be controlled by pressure on either side of the incision or by clamping and tying larger vessels.

*B. Burns, *The Mammalian Cerebral Cortex* [London: Edward Arnold (Publishers) Ltd., 1958].

3. Using a blunt instrument, such as the scalpel handle, separate the skin from the skull and reflect back to expose the skull. Select an area for trephine well forward on the left and away from the midline (to prevent cutting into the superior sagittal veins). The motor cortex of the rat lies well forward in the skull and is very poorly developed compared to the primate. With a dental drill or trephine, cut away the skull to expose the brain. Bleeding can be controlled by small pledgets of cotton or by packing the region with small pieces of bone wax. When the area is large enough, pick up the dura with forceps and carefully expose the surface of the cortex. Keep the exposed area moist with Ringer's solution.

4. When the animal is ready for mapping, a cradle formed from cotton toweling is made, and the rat suspended from stands so that the head, legs, and tail hang free.

5. Stimulate the cortex with platinum electrodes using the inductorium set for moderate strength multiple stimuli. Do not stimulate for long periods or at supramaximal stimulus strengths.

6. Determine the motor control points by noting the muscular response to stimulation of specific points on the cortex. All students should see the point of stimulation and the corresponding muscle action. Referring to Fig. 49 of the rat brain, label the several points successfully stimulated on the rat. When you have completed this portion, you should have a crude but reliable cortex map of the major motor areas. It will be useful to consult the primate map in any advanced physiology text and compare with your rat cortex map.

The following functions should be detectable for your record:

1. **Hind leg motion** – area 4 (primate) via the crossed fibers of corticospinal path, anterior horn cells in the lumbar and sacral region, lumbar and sacral plexi, to muscles of hind limb opposite to side stimulated.

2. **Front leg and shoulder motion** – area 4 via the crossed fibers of corticospinal path, anterior horn cell in cervical and thoracic region, brachial plexus to muscles of shoulder and arm opposite to the side stimulated.

3. **Vibrissa twitch** – area 4 via medial corticobulbar tract, motor nucleus of the facial nerve (VII) in the pons, buccal branch to muscles of the upper lip and vibrissa on same side stimulated, or bilaterally.

4. **Pinna (ear flick)** – area 4 via medial corticobulbar tract, motor nucleus of the facial nerve, posterior cervical branch to muscles of the ear on the same side stimulated, or bilaterally.

At the end of the period, make sure the rat is dead by overetherization and place in the wax bag provided before discarding in waste can.

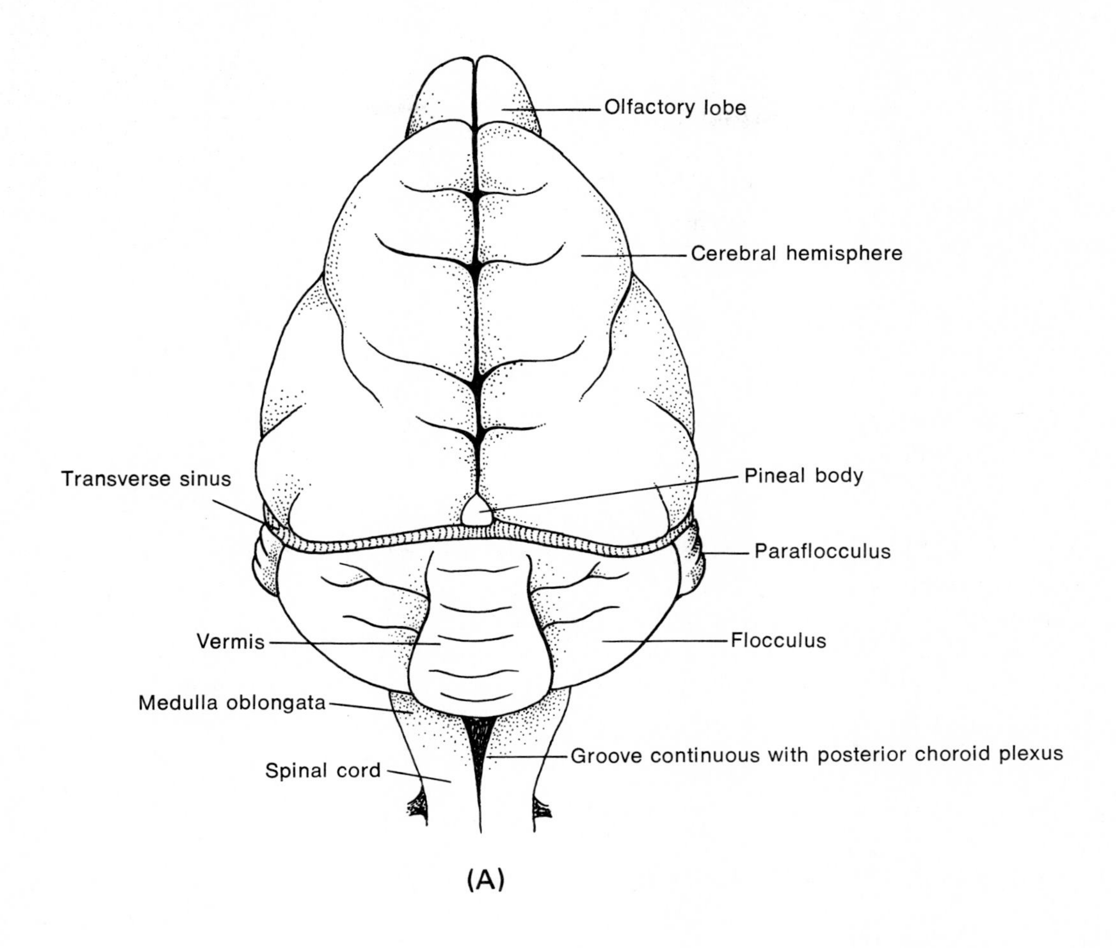

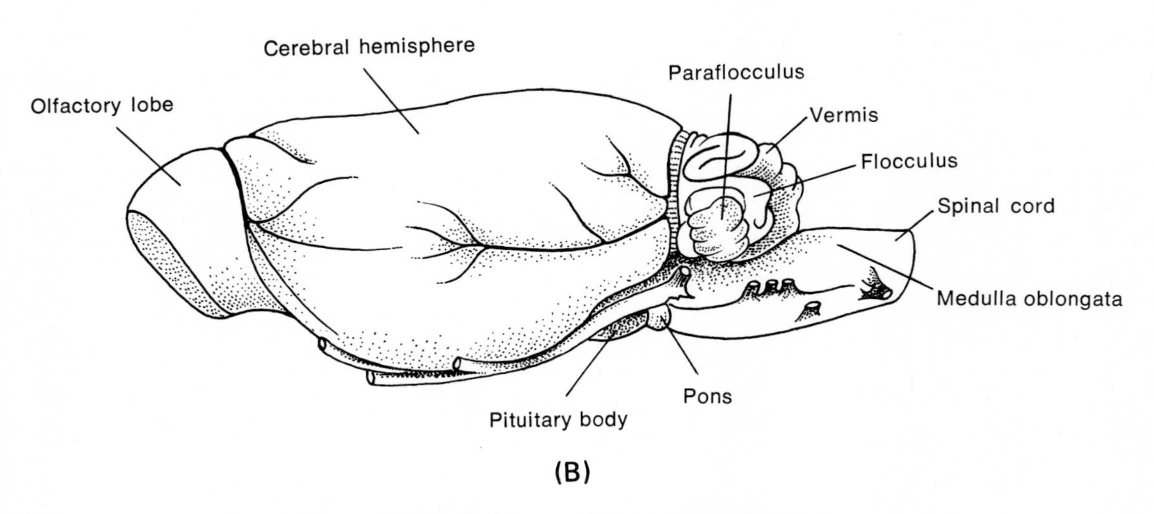

Fig. 49. Rat brain for mapping motor cortical areas.
A. Dorsal view.
B. Lateral view.

Experiment XV

Reflexes in the Human Being

PROCEDURE

1. **Tendon (myotatic or stretch) reflexes**

   a. **Patellar reflex.** Sit on a table so that the leg from the knee down hangs free. The weight of the lower leg puts the quadriceps femoris under tension. Let another student strike the patellar ligament (just below the knee) with the edge of his hand or a ruler. This may require a little practice, and it is best for the subject to divert his attention. Notice the degree to which the leg is extended by the contraction of the quadriceps. Try this on three or four subjects. Is the reflex obtained just as readily and is it equally extensive in all subjects?

   What other name is given to this reflex? Where is the reflex center located? Apart from the strength of the stimulus, in normal individuals by what is the extent of the reflex largely determined? In what two pathologic conditions is this reflex absent or greatly reduced? Why? In what condition is it increased?

   Just before striking the ligament let the subject clench his fist. What effect does this have on the knee jerk? Also give the subject a column of figures to add, and tap the ligament while he is thus engaged. Any reinforcement of the reflex? This is generally attributed to the release from cerebral inhibition.

   b. **Achilles, or ankle, jerk.** Let the subject kneel on a chair, the feet hanging free over the edge of the chair. Bend his foot so as to increase the tension of the gastrocnemius muscle. Tap the tendon of Achilles. Result? This is the tendon of what muscle? The contraction of this muscle causes plantar flexion of the foot. Feel the muscle during the tapping.

2. **Cutaneous (superficial) reflexes**

   a. **Corneal reflex.** Touch the cornea with a soft object. What afferent and what efferent nerves are concerned? Where is the center located? What is the "object" of this reflex?

3. **Organic reflexes**

a. **Photopupil reflex.** Let the subject close the eyes for 2 min; while facing a bright light, he should open them and let another student examine the pupils immediately. What muscle causes this action? Name the afferent and efferent nerves. Where does the afferent nerve originate? Where is the reflex center located? What is the object of this reflex?

b. **Accommodation pupil reflex.** In a moderate light let the subject look at a distant object (20 or more feet removed) and examine the pupil. Now have him look at a pencil held about 10 in. from the face (without changing the illumination) and note pupil. Object of this reflex?

c. **Ciliospinal reflex.** Pinch the skin on the nape of the neck and note the dilation of the pupil.

d. **Convergence reflex.** While the subject is looking at a distant object, note the position of the eyeballs. Let the subject look at a near object. What change occurs in the eyeballs? What muscle brings this about? What is the motor nerve? What is the object of this convergence? Do you know of any abnormal condition in which this action is lacking? What is the result of this?

e. **Swallowing reflex.** Swallow the saliva in your mouth and immediately swallow again. You will find that it is impossible to do so if there is no saliva to swallow the second time. That this is not because swallowing acts cannot be performed in rapid succession is demonstrated by rapidly drinking a beaker of water. Explain. Is swallowing a volitional act?

Experiment XVI

Afferent and Efferent Nerve Fibers

APPARATUS

Stimulating apparatus, frog board, thread.

PROCEDURE

1. Pith the brain of a frog; place the frog on a frog board (dorsum down); open the abdomen and remove the viscera. This exposes the origin of the sciatic nerve (three roots). With a glass hook gently lift up the three nerves and apply a loose ligature that leaves the nerves uninjured, but by which the exposed nerve can be readily and gently lifted onto the electrodes. Also place a loose ligature under the roots of the right sciatic nerve. Remove the skin from both legs, so that feeble twitching of the muscles can be observed.

2. Place the left roots onto the platinum electrodes, making sure that the electrodes are not in contact with the neighboring tissues. For 1 sec stimulate with a strength of interrupted faradic current that will cause contraction of the homolateral muscles only. So far as you were able to tell, what fibers in the mixed sciatic nerve were stimulated and conveying impulses?

3. Now increase the strength of the current considerably (1-sec stimulation). Results are now seen in both legs. What fibers, in addition to those used above, were stimulated? How did the impulses in these fibers reach the contralateral leg? What structures are necessary for this?

4. Repeat step 3, using the right sciatic nerve.

5. In the middle of the thigh place two ligatures (about 5 mm apart) on the right sciatic nerve (tie them securely). With scissors cut the nerve between the ligatures. Stimulate the left nerve with the current that gave good results in step 3.

6. Stimulate the peripheral end of the cut nerve and watch for both homolateral and contralateral results. Now stimulate the central end.

Draw a diagram you can utilize in discussing the results of this experiment.

Experiment XVII | Muscle Tonus

APPARATUS

Stand, clamp, Ringer's solution, dropper

PROCEDURE

1. Cut off the upper jaw of a frog; open the abdomen, remove viscera, and expose the roots of the left sciatic nerve. Suspend the frog by a hook passed through the lower jaw. After shock has disappeared, note the characteristic "bends" of the hind legs at the hip, knee, ankle, toes. Make a line drawing of this. Also note that the long toes of the two legs hang equally low.

2. Cut all the roots of the right sciatic and repeat the observation outlined in step 1. Account for the difference in the two legs. This difference is sometimes seen better if the frog is suspended in a deep glass vessel filled with water.

3. Expose both gastrocnemius muscles and, while the frog is suspended, compare the firmness of these two muscles. Why this difference?

4. Destroy the spinal cord; repeat observations made in steps 1 and 3.

5. Take note of your lower jaw. Do you need to exert any effort to keep it raised? Why not? Now lower the jaw for a few seconds. Is the feeling as comfortable as when the jaw was allowed to be raised? What happens as soon as your attention is distracted? Why?

6. What kind of action is muscle tonus? What is the origin of the afferent impulses? What kind of reflex is this sometimes called? In the human body muscle tonus may be aided by impulses originating in what other sensory surfaces? Of what value is muscle tonus to us? Which muscles in the body are especially kept in tonus? What are these muscles sometimes called? During what normal, recurring condition is tonus largely lost? What is the evidence for this?

Experiment XVIII | Reaction Time

APPARATUS

Kymograph, two keys (see instructions), cells, signal magnet, tuning fork.

Because the speed of the drum must be uniform (unless an electric tuning fork is provided), it is best to wind the kymograph to the end before each tracing is made. The object of the experiment is to determine the length of the reaction time to sight, sound, and touch.

PROCEDURE

1. Set up the apparatus as shown in demonstration in lab. The subject whose reaction time is to be determined operates the key marked A; the investigator operates key B. The subject closes his key, and when the investigator also closes his key, the circuit is complete and the signal magnet will draw the pen down. If now the subject opens his key, the circuit is broken and the pen goes up. Try it. Use fastest speed of drum, and after the tracing has been made, wind the drum. Also start the drum revolving at the same point with respect to the pen of the signal magnet (why?). Then make a tuning fork tracing below the reaction time tracing.

2. Reaction time to sight. The subject closes key A and while the drum is rotating watches for the pen of the signal magnet to move. The investigator closes key B, and as soon as the subject perceives this, he opens key A. The length of the break in the horizontal line written by the signal magnet, in terms of the vibration of the tuning fork, indicates the reaction time to visualize stimulation. Repeat this at least three times and take the average.

3. Reaction time for sound. For key B substitute a spring key, the closure of which is audible. Repeat as in step 2, but the subject, instead of looking for the signal, listens for the click of key B.

4. Reaction time to tactile stimulation. Key B must be a "silent" key. Cut the wire between the two keys; the subject takes the two ends between his finger and thumb. When he feels the shock caused by the closing of key B, he opens key A and thus records his reaction time to touch.

5. Tabulate your results. What various processes contribute to the formation of the reaction time? How can you explain that the reaction time for sound is less than for sight?

33

Circulatory System

The anatomy of the circulatory system appears to be one of the most difficult phases in the dissection of the cat. Care must be used in tracing out the circulatory system, since it is rather easy to cut and damage the vessels, making the tracing out of the vessels practically impossible. The cats used in the laboratory are injected with Latex compound. The veins are blue and the arteries are red. Microscopic sections of an artery, vein, and capillary are seen in Figs. 50, 51, and 52. Note the difference between the cellular layer structure of these vessels.

The Heart (Sheep) (Figs. 53 and 54)

In the laboratory the heart of the sheep is usually used for study because it is approximately the size of the human heart. Remove as much fat and adhering connective tissue from the surface of the heart as you can and place the heart with its ventral surface up and with the apex pointed toward you. Use the cat dissection to orient yourself properly.

Right and left sides of the heart can be determined by the fact that the left ventricle includes all the **apex** of the heart, which is the more pointed part and is directed toward the diaphragm. The left side of the heart feels more firm and muscular than the right side, which feels soft and flabby. The left side of the heart represents from the outer surface the location of the **left ventricle**, and its right side, **the right ventricle.** The **base** of the heart is the flat upper portion of the heart, from which project several large blood vessels, and is mainly formed by the right and left **atria.** The **atria** appear to be concentric masses separated from each other by the pulmonary artery and the aorta. The atria are separated from the much larger ventricles by the **coronary sulcus** (groove), which is continuous around the whole heart until the pulmonary artery is reached.

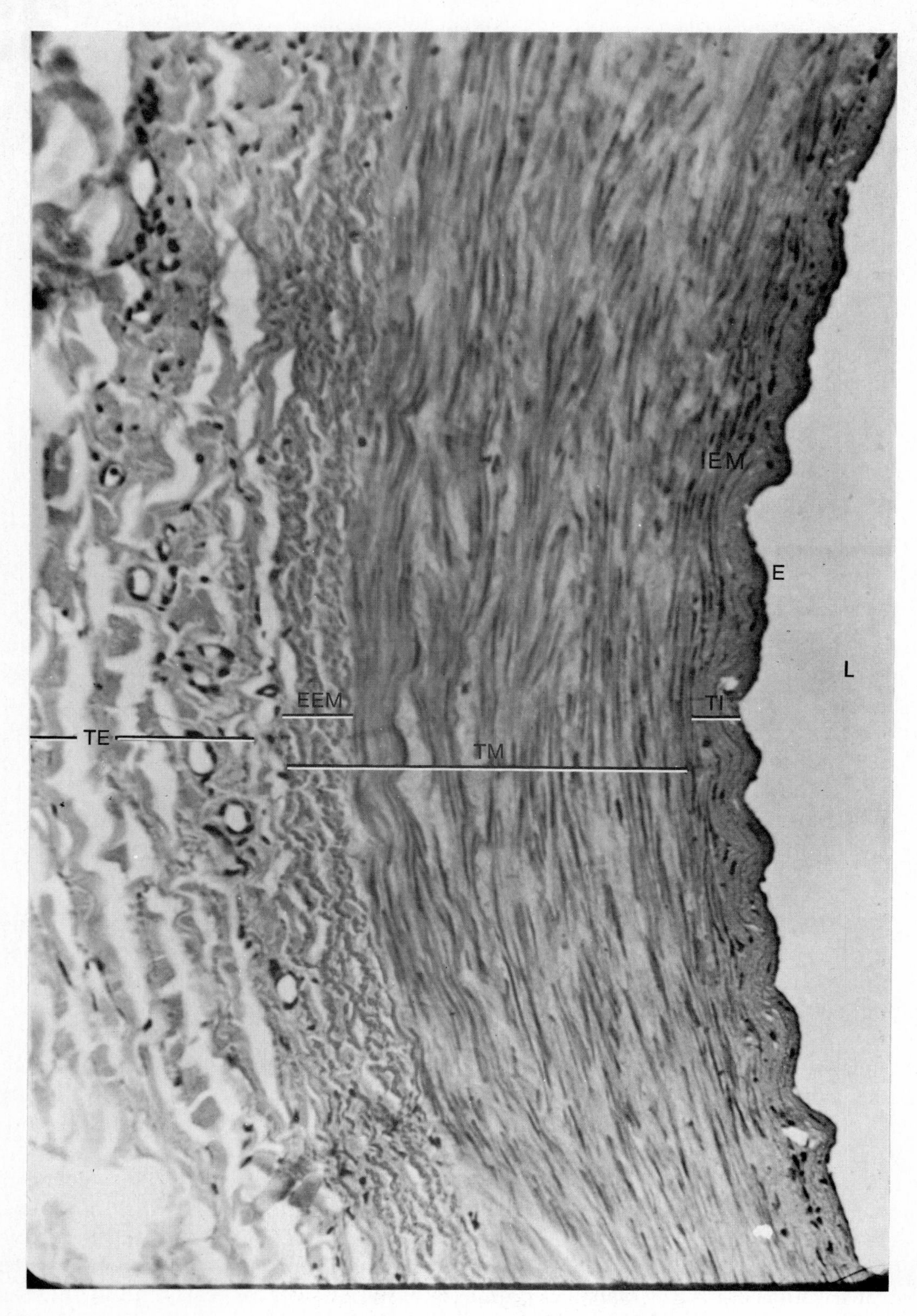

Fig. 50. Cross section of artery. L, lumen; E, endothelium; IEM, internal elastic membrane; TI, tunica intima; TM, tunica media; EEM, external elastic membrane; TE, tunica externa.

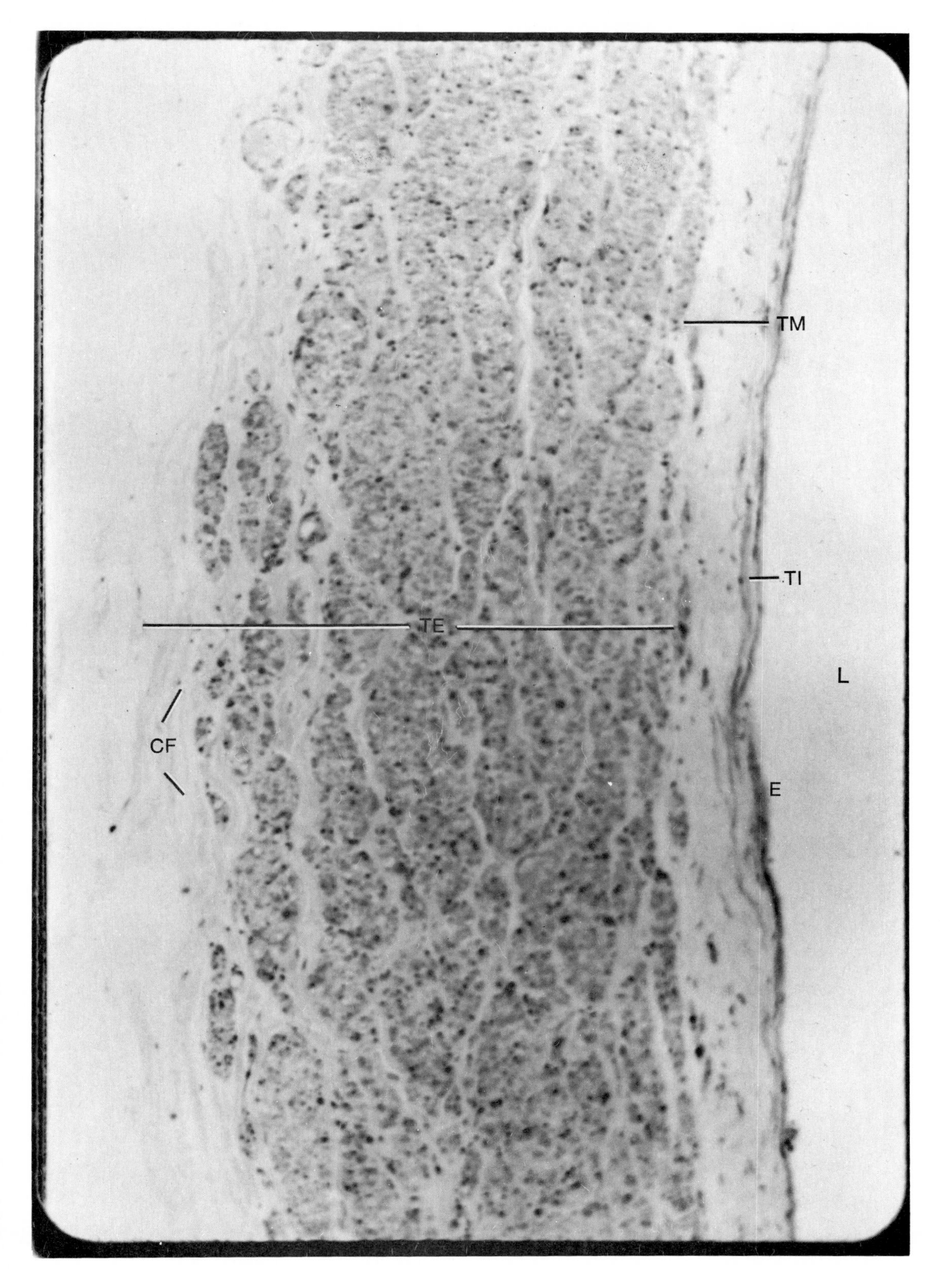

Fig. 51. Cross section of vein. L, lumen of vein; E, endothelium; TI, tunica interna; TM, tunica media; TE, tunica externa (aduentitia); CF, collagenous fibers.

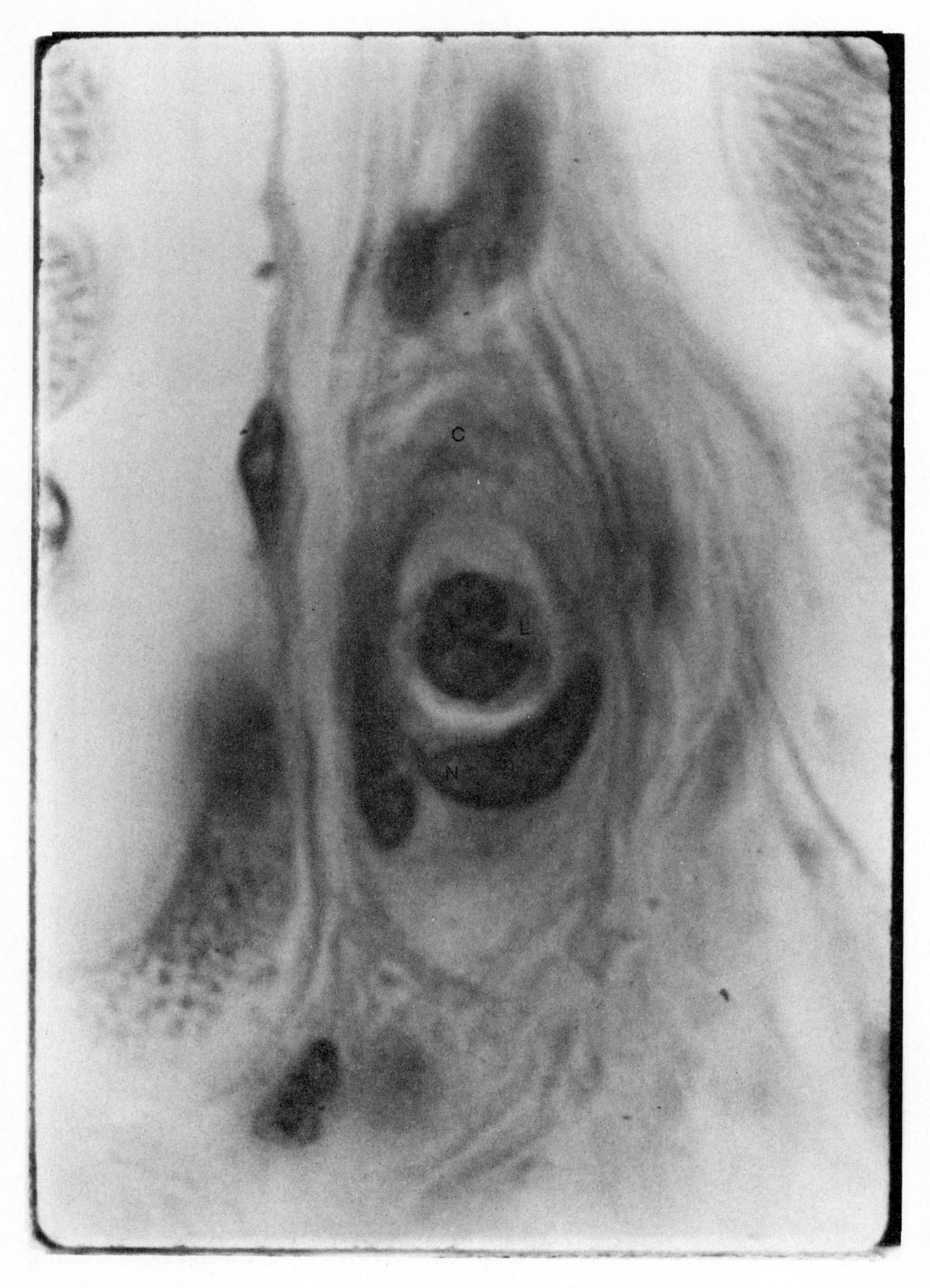

Fig. 52. Cross section of capillary. C, capillary wall; N, nucleus of endothelial cell; L, lymphocyte in lumen of capillary.

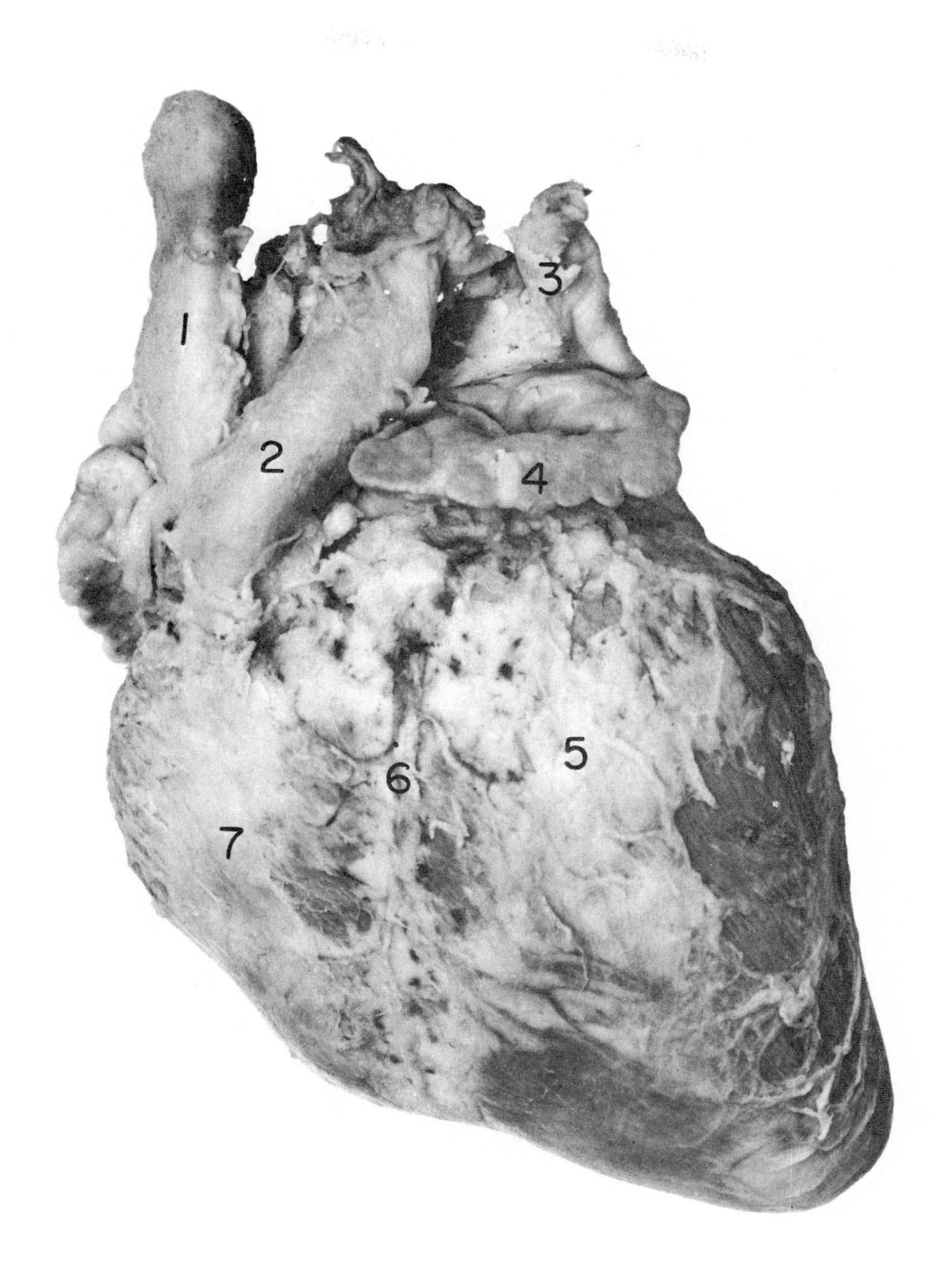

Fig. 53. Sheep's heart, left side.

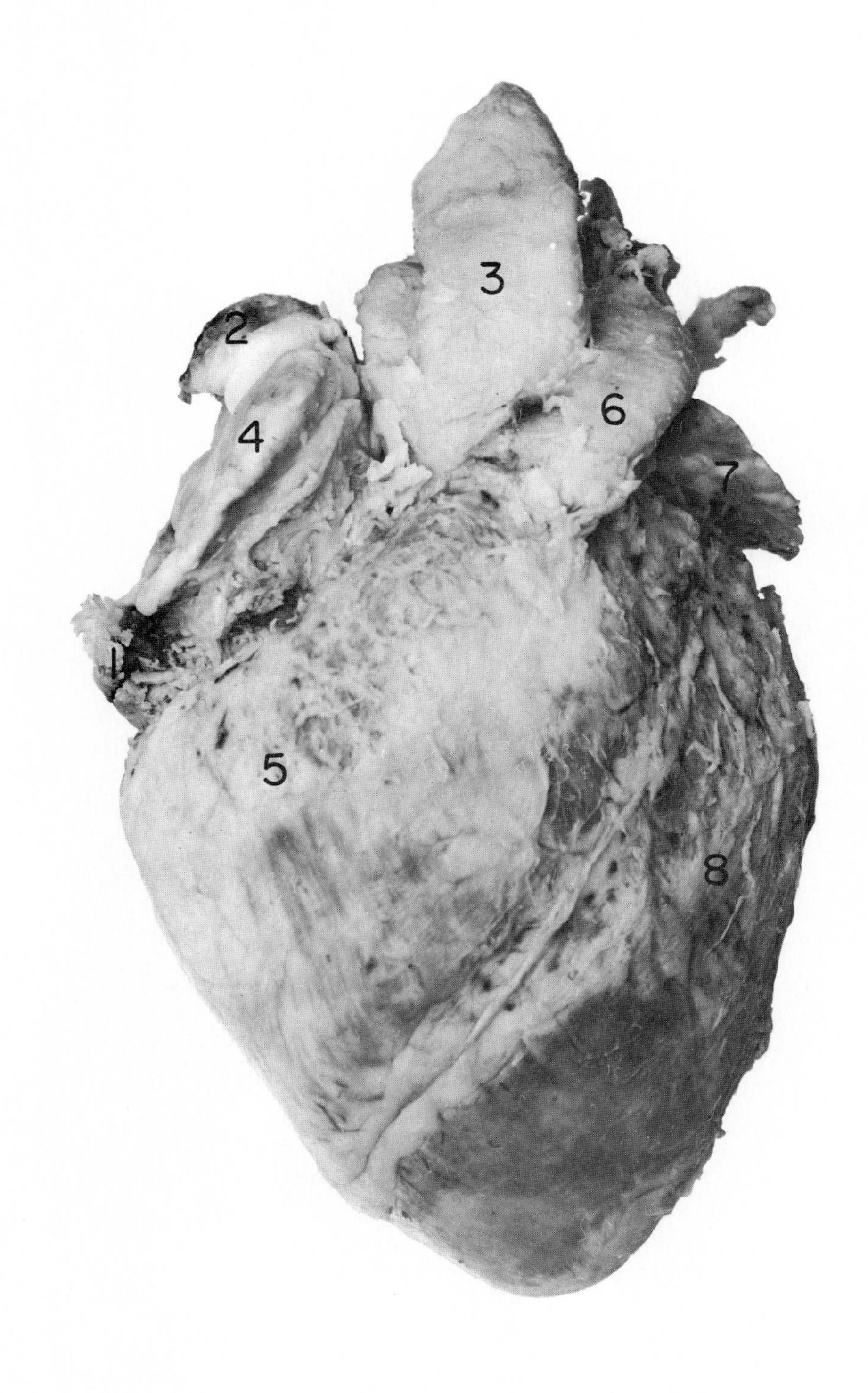

Fig. 54. Sheep's heart, right side.

STRUCTURES TO BE IDENTIFIED ON THE EXTERNAL SURFACE OF THE HEART

Left side of the heart (Fig. 53)

1. Aorta
2. Pulmonary artery
3. Pulmonary veins
4. Left atrium
5. Left ventricle
6. Coronary artery
7. Right ventricle

Right side of the heart (Fig. 54)

1. Inferior vena cava
2. Superior vena cava
3. Aorta
4. Right atrium
5. Right ventricle
6. Pulmonary artery
7. Left atrium
8. Left ventricle

INTERIOR OF THE HEART

Open the right atrium with your scissors by making an incision from the termination of the superior vena cava down to a point where the inferior vena cava joins the heart. From the middle of this incision cut through the wall of the atrium and well down into the right ventricle. Wash the interior with tap water and remove any blood clots that may be present.

Identify the following (Fig. 55) (see text: Grollman's *The Human Body: Its Structure and Function)*:

1. **Opening of inferior vena cava** – a probe passed through this structure will show where it enters.
2. **Opening of superior vena cava.**
3. **Fossa ovalis** – a deep oval depression between the atrial septa, directly across from the entrance of the inferior vena cava.
4. **Coronary sinus** – slitlike opening below the fossa ovalis.

Venous blood is carried into the right atrium by three vessels: the two vena cavae and the coronary sinus. Blood passes from the atrium into the right ventricle by a large rounded **atrioventricular** orifice, which occupies the whole of the floor of the right atrium.

RIGHT VENTRICLE

Make your incision into the right ventricle parallel to the coronary sulcus and toward the pulmonary artery. Near the pulmonary artery continue the incision of the right ventricle parallel to and a short distance from the longitudinal groove, back to the right border of the heart.

Identify the following (Fig. 55):

1. **Tricuspid valve** – three triangular membranous cusps in the opening between the atrium and the ventricle.
2. **Trabeculae carneae** – fleshy ridges on interior wall.
3. **Papillary muscles** – three or more conical muscular processes to which are attached tendinous chords, **the chordae tendineae**, which run across the ventricle to attach to the tricuspid valves.
4. **Pulmonary orifice** – opening of ventricle into the pulmonary artery. Note the three pocket shaped **semilunar valves**, which prevent the backward flow of blood from the artery into the ventricle.

LEFT ATRIUM

To open the left atrium, make an incision, beginning far back and parallel to the coronary sulcus, forward to the extremity of the auricular appendage. Wash the cavity with tap water and identify the following:

1. Opening of pulmonary veins – on its posterior wall.
2. Opening of the left atrioventricular orifice.

LEFT VENTRICLE

To open the left ventricle, make an incision, beginning near the coronary sulcus and parallel to one of the longitudinal grooves, around the apex and up to the other side parallel to the other longitudinal groove. Notice that the **trabeculae carneae** are less conspicuous than in the right ventricle, and the papillary muscles are larger. Examine the **mitral valve** or **bicuspid valve** between the atrium and the ventricle. With a probe or finger locate the **aortic orifice**, which is the opening from the ventricle into the aorta. This opening is guarded by the **semilunar valves.**

Arteries

ARTERIES OF THE THORAX AND HEAD REGIONS (Fig. 56A)

Remove any more of the chest wall that is necessary to expose all the blood vessels of the chest, neck, and the side of the head. Locate and study the following:

1. **Aortic arch** – large vessel that forms an arch as it leaves the heart and turns posteriorly. Branches of the aortic arch are the **innominate** and **left subclavian** arteries.
2. **Pulmonary artery** – soon branching into right and left pulmonary arteries as it leaves the heart from the right ventricle.
3. **Innominate artery** – largest artery branching from the aortic arch. Its branches are many, but three are the largest: **right subclavian, left** and **right common carotids.** The branches of these will be considered next.
4. **Left common carotid artery (LCC)** – first branch from the innominate carrying blood to the head. This one will be traced first, and give off the following branches: thyroid arteries, muscular, occipital, internal and external carotids.

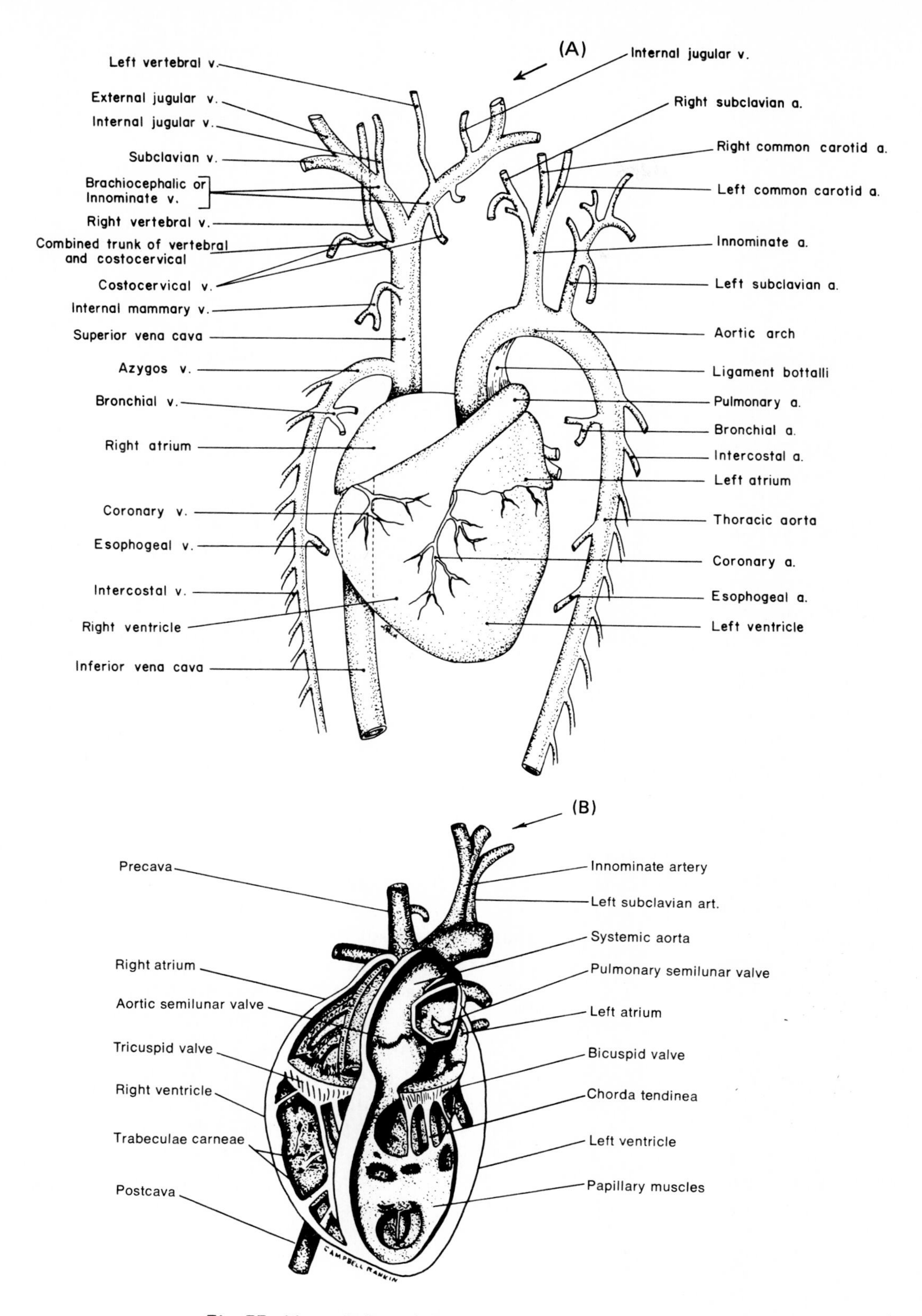

Fig. 55. Ventral view of the cat heart.
A. External surface and main blood vessels with their branches.
B. Dissection.

5. **Superior thyroid artery** – given off of left common carotid at the level of the thyroid gland and supplying it.
6. **Muscular artery** (branch of LCC) – at the same level as the thyroid artery and passing to the back of the neck to supply muscles there.
7. **Occipital artery** (branch of LCC) – slightly anterior to the thyroid; supplies neck muscles.
8. **Internal carotid artery** (branch of LCC) – at the level of the occipital – enters the skull supplying the brain. In the cat this artery is usually not present. If it is, it is quite small.
9. **External carotid artery** (branch of LCC) – anterior to the internal carotid – supplies the face and head; has many branches such as the lingual, auricular, and temporal arteries.
10. **Lingual artery** (from the external carotid) – supplies the tongue.
11. **Auricular artery** (from the external carotid) – supplies the pinna of the ear.
12. **Temporal artery** (from the external carotid) – supplies the temporal region.
13. **Internal maxillary artery** – (branch from temporal) – to the masseter muscle.
14. **External maxillary artery** – to the upper and lower jaws and to the lips.
15. **Right common carotid artery** (branch from innominate) – the branches from this are almost the same as those of the left common carotid and will not be listed. They should be traced out just as were those from the left common carotid.
16. **Right subclavian artery** – this may appear to be a branch from the right common carotid in some specimens. Branches of the right subclavian are the vertebral, sternal, costocervical, thyrocervical and axillary.
17. **Vertebral artery** – given off the subclavian at about the region where that artery leaves the thoracic cavity. It supplies neck muscles by small branches, then passes to the brain. (Some connective tissue may be removed or loosened after lifting the subclavian artery.)
18. **Sternal artery** (or **internal mammary**) – from the ventral side of the subclavian; supplies the ventral body wall.
19. **Costocervical axis** – branch from the subclavian just opposite the first rib. It later branches into two arteries supplying the back and neck.
20. **Thyrocervical axis** – rises from the subclavian below the first rib. It branches to supply the neck and shoulder.
21. **Axillary artery** – the continuation of the subclavian into the front leg and out of the thoracic cavity. Branches of the axillary artery include the ventral thoracic, long thoracic, subscapular, and brachial arteries.
22. **Ventral thoracic artery** (from the axillary) – supplies the pectoral muscles.
23. **Long thoracic artery** (from the axillary) – supplies the latissimus dorsi muscle.
24. **Subscapular artery** (from the axillary) – divides into several branches; the most important being the **thoracodorsal artery**, which supplies the latissimus dorsi and other nearby muscles.
25. **Brachial artery** – a continuation of the axillary artery beyond the point of branching of the subscapular. It gives off many branches to the foreleg muscles above the elbow.
26. **Radial artery** – a continuation of the brachial artery below the elbow. It gives off branches to the lower front leg and to the toes.
27. **Left subclavian artery** (branch from aortic arch) – gives off similar branches to those of the right subclavian. It should be traced to fix the arteries firmly in mind.

Branches of the Aorta in the Thoracic Cavity

29. **Intercostal arteries** – ten pairs of them supply the thoracic walls.
30. **Bronchial arteries** – emerge from the thoracic aorta opposite the fourth intercostal space or from the fourth intercostal arteries – supply the bronchi.
31. **Esophageal arteries** – supply the esophagus. Small branch from right common carotid and also from aorta – supply the esophagus.

ARTERIES OF THE ABDOMINAL REGION (Fig. 56B)

Trace the aorta through the thoracic cavity to the diaphragm. Notice that it passes into the abdominal cavity. Identify the following arteries in the abdominal region: The number beside each artery represents its location in Fig. 56B, page 180.

1. **Descending aorta** *(abdominal aorta)*
2. **Splenic artery** (branch of celiac) – supplies the spleen, the pancreas, the greater curvature of the stomach and the omentum.
3. **Celiac artery** – first major branch from the abdominal aorta just posterior to the diaphragm. It gives off the following three branches (hepatic, left gastric, and splenic).
4. **Left gastric artery** (branch of celiac) – supplies the stomach.
5. **Hepatic artery** (branch of celiac) – supplies the liver. It will be necessary to separate the stomach and liver to find this artery.
6. **Superior mesenteric artery** – branch of the abdominal aorta just posterior to the celiac artery; supplies the intestines and the pancreas. It has several branches that will be considered next.
7. **Middle colic artery** – branch of the superior mesenteric; supplies the transverse and descending colon.
8. **Inferior pancreaticoduodenal artery** – branch of the superior mesenteric; supplies the duodenum and the pancreas.
9. **Right colic artery** – branch of the superior mesenteric; supplies the ascending colon.
10. **Iliocolic artery** – branch of the superior mesenteric; supplies the cecum and the lower ileum.
11. **Intestinal arteries** – branches of the superior mesenteric (many of these); supply the intestines.
12. **Adrenolumbar arteries** (a pair of them) – arise from the aorta just posterior to the superior mesenteric artery and supply the adrenal glands and the back.
13. **Phrenic artery** – branches from each adrenolumbar artery supplying the diaphragm. *(It may branch from the celiac artery instead.)*
14. **Renal arteries** (a pair of them) – branch from the aorta to the kidneys.
15. **Ovarian arteries** (in the female) or the **spermatic arteries** (in the male) – branch next from the aorta. The ovarian arteries supply the ovaries, and the spermatic arteries supply the testes. Sometimes they branch from the renal arteries.
16. **Lumbar arteries** – occasional arteries from the aorta to the back muscles.
17. **Inferior mesenteric artery** – large single branch from the aorta just posterior to the genital arteries, supplies the lower colon and the rectum. It branches into the next two arteries.
18. **Left colic artery** – branch from the inferior mesenteric; supplies the descending colon.
19. **Superior hemorrhoidal artery** – branch from the inferior mesenteric; supplies the rectum.
20. **Iliolumbar arteries** (a pair of them) – branch from the aorta posterior to the inferior mesenteric and supply the psoas and the iliopsoas muscles.
21. **External iliac arteries** (a pair of them) – branch from the aorta at the posterior end of the abdominal cavity. Be careful not to injure the urogenital organs while looking for these arteries. They divide into two branches and supply the hind legs.
22. **Deep femoral artery** – branch of the external iliac. It branches to send the **inferior epigastric artery** to the **abdomen** and other branches to the genital organs and the bladder.
23. **Femoral artery** – branch of the external iliac; supplies the leg. It also branches in the upper thigh laterally into the **anterior femoral artery.**
24. **Internal iliac arteries** (a pair of them) – branch posterior to the external iliacs. Each one gives off several branches.

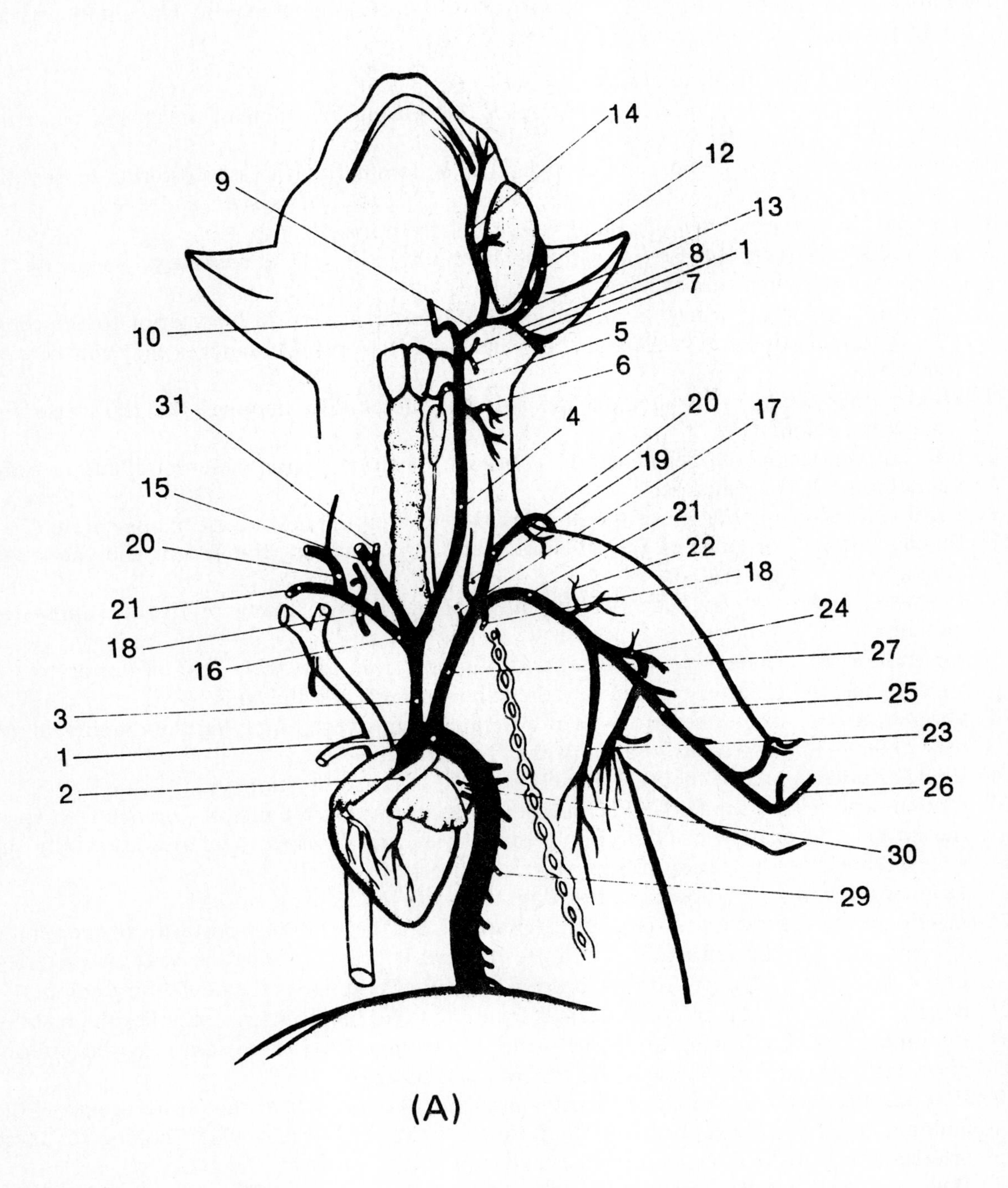

Fig. 56A. Arteries of the thorax and head. The numbers correspond to those on pages 174-176.

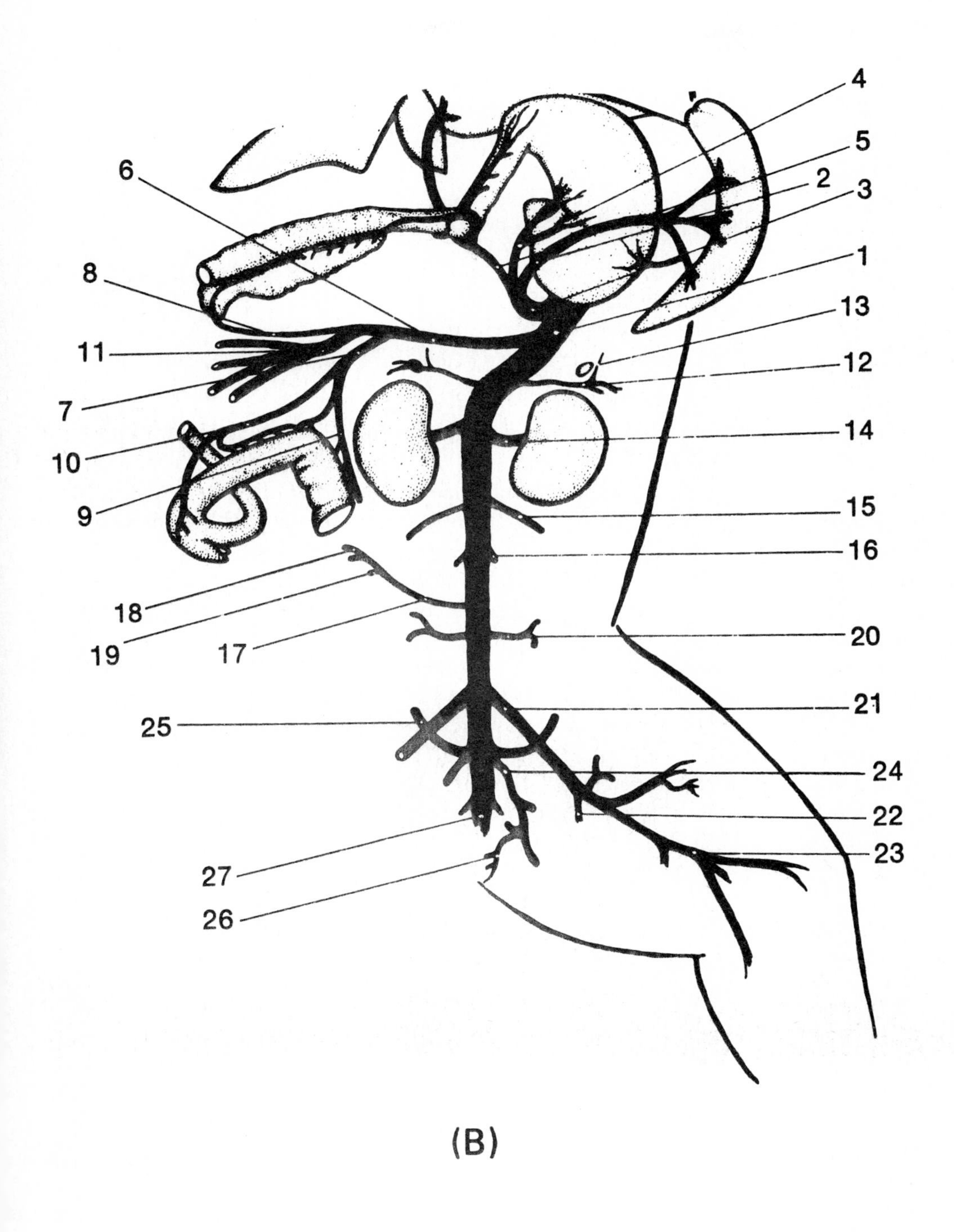

Fig. 56B. Arteries of the abdominal region. The numbers correspond to those on pages 177-179.

25. **Umbilical artery** – branch of the internal iliac; supplies the bladder.
26. **Middle hemorrhoidal artery** – branch of the internal iliac posterior to the umbilical; supplies the rectum. The **uterine artery** branches from the middle hemorrhoidal in the female.
27. **Caudal artery** (or **sacral artery**) – the continuation of the descending aorta posterior to the branches mentioned above; supplies the sacrum and the tail.

Veins

VEINS OF THE THORAX AND THE HEAD (Fig. 57A)

Very little dissection will be required to expose the veins since they lie along the same regions as do the arteries. Begin at the heart and locate and study the following:

Veins Entering The Heart. These are the superior vena cava, inferior vena cava, and pulmonary veins.

1. **Precaval vein** (or **superior vena cava**) – joins the right atrium. The next four veins will be branches of this large vein.
2. **Azygos vein** – just anterior to the point of entrance of the precaval vein into the heart. Trace this vein and find the intercostal veins from between the ribs, the bronchial veins from the bronchi, and the esophageal veins from the esophagus.
3. **Internal mammary veins** – a pair, but united at the root which enters the precaval vein opposite the third rib. They continue through the diaphragm and become the **superior epigastric** veins in the abdomen.
4. **Right vertebral vein** – enters dorsal surface of precaval vein by a common trunk, formed by a union with the costocervical vein. It draws blood from head region.
5. **Brachiocephalic** or **innominate veins** – unite at the anterior margin of the thoracic cavity to form the precaval vein.
6. **Left vertebral vein** – enters the dorsal surface of the left brachiocephalic vein.
7. **Costocervical vein** – enters the left vertebral vein just prior to the entrance of the vertebral into the brachiocephalic. It is deeply embedded as it receives branches from the back and neck muscles.

Veins Draining Into Right Brachiocephalic (Innominate). These are the external jugular and subclavian veins.

8. **External jugular veins** – draws blood from the head and joins the brachiocephalic along with the subclavian vein. These two now extend in different directions.
9. **Subclavian vein** – becomes the axillary after it passes through the body wall. Subclavian receives blood from the axillary and subscapular veins.
10. **Axillary vein** – the continuation of the subclavian vein in the shoulder region.
11. **Subscapular vein** – the first large branch of the axillary vein from the upper arm and shoulder.
12. **Ventral thoracic vein** – enters the axillary laterally from the subscapular. It serves the pectoral muscles.
13. **Long thoracic vein** – just beyond the ventral thoracic. It also comes from the pectoral muscles.

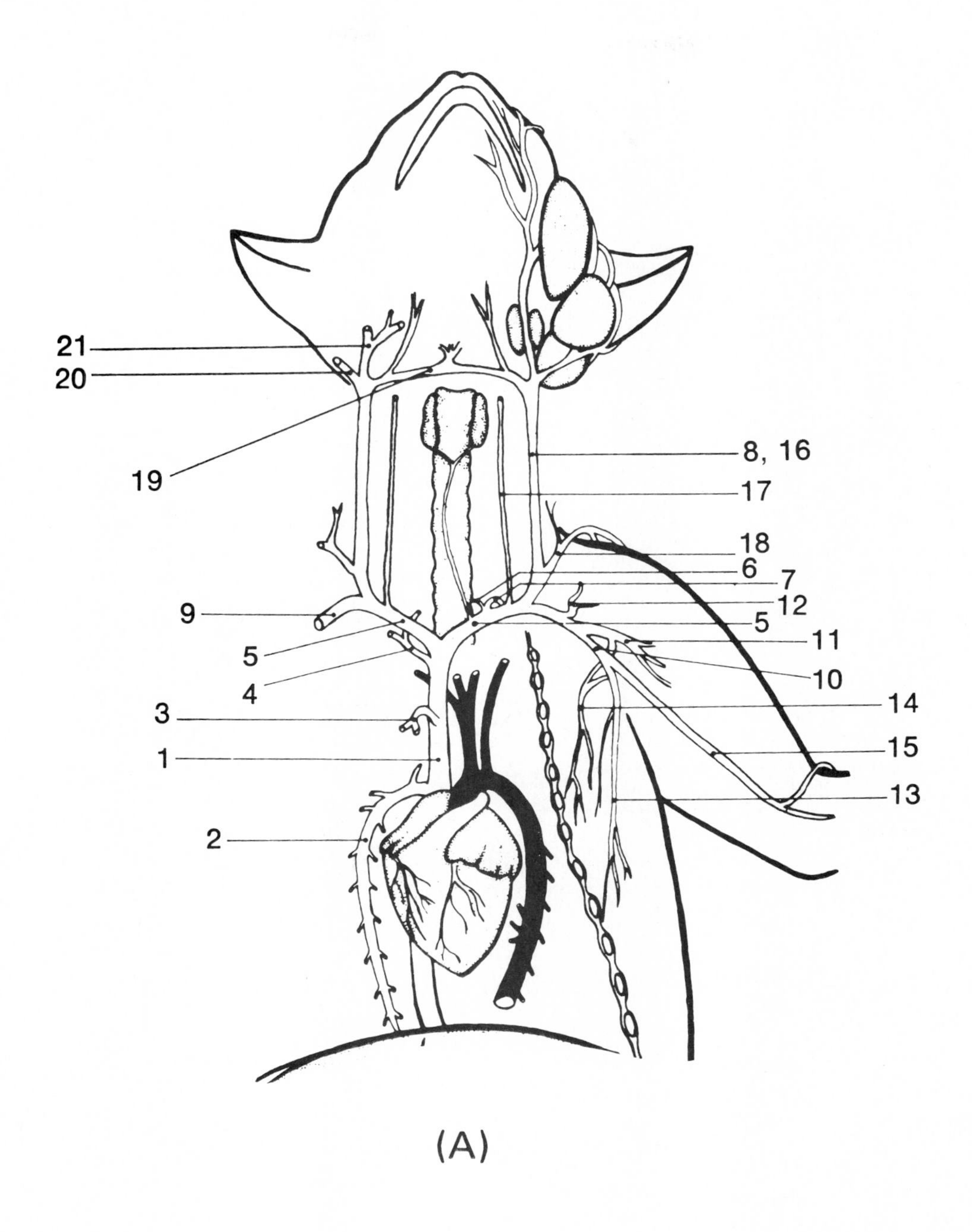

Fig. 57A. Veins of the thorax and head. The numbers correspond to those on pages 179 and 181.

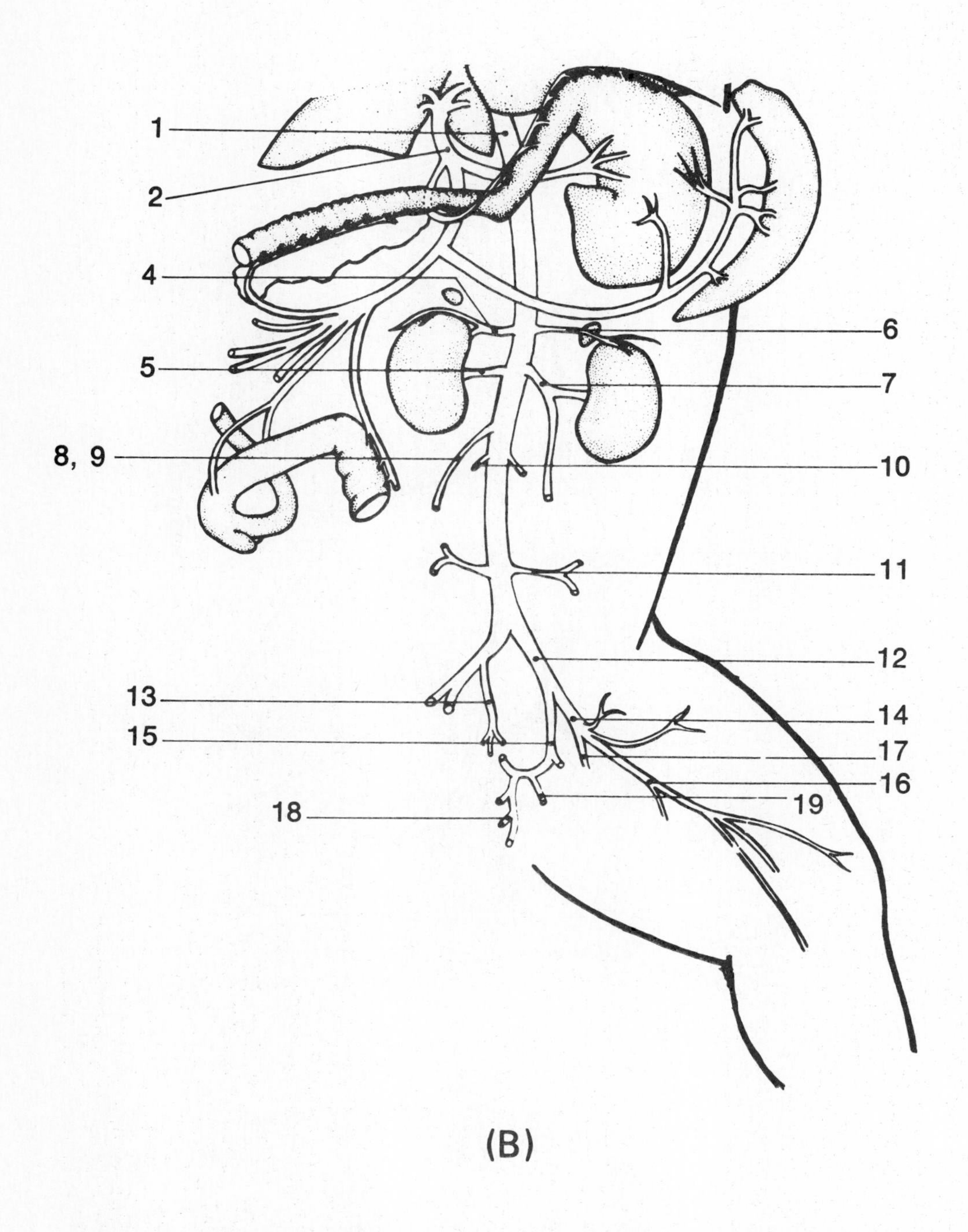

Fig. 57B. Veins of the abdominal region. The numbers correspond to those on page 181-182. Numbers 2 (phrenic) and 3 (hepatic) are not shown because they leave from the dorsal surface and cannot be seen in this view.

14. **Thoracodorsal vein** – joins the axillary near the long thoracic vein and draws blood from the latissimus dorsi muscle.
15. **Brachial vein** – the continuation of the axillary vein into the arm.
16. **External jugular vein** (paired) – the other large branch from the brachiocephalic. Trace it toward the head. The next five veins are branches from this vein.

Veins Draining Into External Jugular

17. **Internal jugular vein** – enters the external jugular near its junction with the subclavian. Drains blood from the head and the brain.
18. **Transverse scapular vein** – joins the external jugular at the base of the neck. It carries blood away from the shoulder and the upper arm.
19. **Transverse jugular** – unites the two jugulars at the base of the chin.
20. **Posterior facial vein** – enters the external jugular near the juncture of the transverse jugular vein. It in turn branches into the posterior auricular vein, the anterior auricular vein, and the superficial temporal vein. These last two unite before entering the posterior facial. The posterior facial receives blood from the ear, parotid gland, and maxillary area.
21. **Anterior facial vein** – enters the external jugular receiving blood from the face, jaws, and the submaxillary gland.

    Thoracic duct – great vessel of lymphatic system entering the external jugular at the point where the jugular enters the subclavian.

VEINS OF THE ABDOMINAL CAVITY (Fig. 57B)

Very little dissection will be needed here, but be careful not to disturb the urogenital organs and the nerves. Begin at the heart and proceed toward the posterior. Locate the following:

1. **Postcaval vein** (inferior vena cava) – enters the right atrium.
2. **Portal vein** – formed by union of the inferior mesenteric vein, the superior mesenteric vein, and the splenic vein to liver (Fig. 58).
3. **Hepatic veins** – leaving the liver, may be seen by dissecting out the liver around the post-caval vein out of liver to postcaval.
4. **Right adrenolumbar vein** – first branch of the postcaval posterior to the liver. It serves the right adrenal gland and inner part of the back.
5. **Right renal vein** – from the kidney, enters the postcaval just posterior to the right adrenolumbar.
6. **Left adrenolumbar** – slightly posterior to the right renal; serves the left adrenal gland.
7. **Left renal vein** – slightly posterior to the left adrenolumbar; serves the left kidney. (The left spermatic vein of male cats enters the left renal; the left ovarian vein of female cats enters the left renal.)
8. **Right internal spermatic vein** – enters the postcaval vein slightly to the posterior to the kidneys (in male cats).
9. **Right ovarian vein** – enters the postcaval just posterior to the kidney (in female cats).
10. **Lumbar veins** – a number of pairs of these enter the postcaval at varying intervals. Lift the vein carefully to find these, for they enter on the dorsal side.
11. **Iliolumbar veins** (paired) – enter the postcaval in the groin region. They are large and are accompanied by the corresponding arteries.
12. **Common iliac veins** (paired) – join just posterior to the iliolumbar to form the postcaval.
13. **Caudal vein** – a branch from the left common iliac; serves the tail.

14. **External iliac vein** – the common iliac is formed by the union of this vein with the internal iliac.
15. **Internal iliac vein** – joins the external iliac to form the common iliac.
16. **Femoral vein** – from the leg, joins with the deep femoral to form the external iliac vein.
17. **Deep femoral vein** – joins with the femoral to form the external iliac just posterior to the entrance of the internal iliac.
18. **Middle hemorrhoidal vein** – joins the internal iliac from the rectum and the bladder.
19. **Gluteal veins** – join the internal iliac from the hip region.

PORTAL CIRCULATION (Fig. 58).

The portal vein receives blood from the large and small intestines, stomach, spleen, and pancreas and conveys it to the liver. The portal blood is distributed within the liver by a set of venous capillaries that empty into the hepatic veins, from which the blood passes into the inferior vena cava. The portal blood carries absorbed food material from the intestine to the liver.

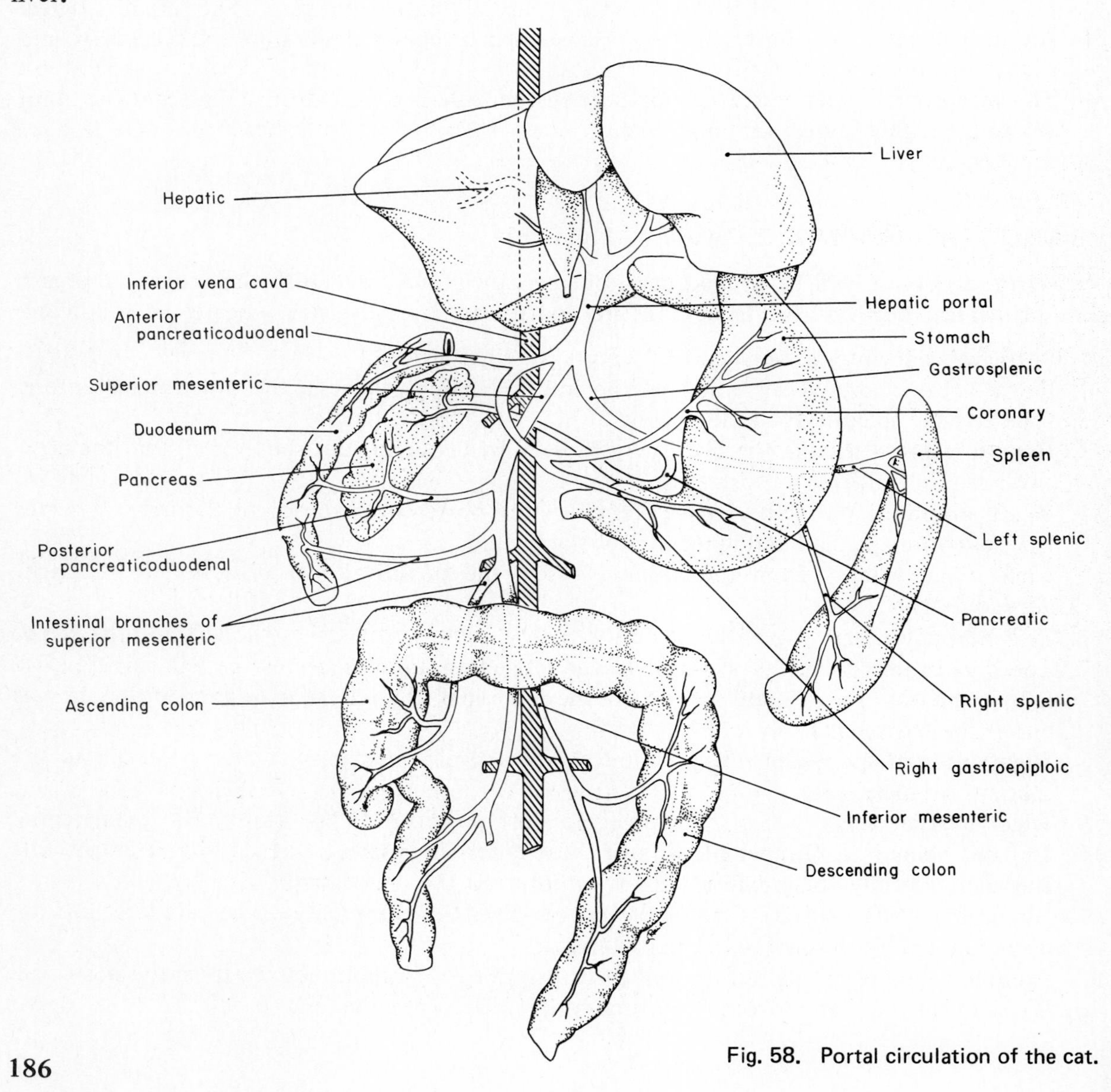

Fig. 58. Portal circulation of the cat.

34

Digestive System

The digestive system consists of a long, muscular tube into which the food is taken and within which the digestive processes occur. The tract can be divided into several portions: the **mouth, pharynx, esophagus, stomach, small intestine, large intestine**, and **rectum.** Several accessory structures associated with the digestive system perform important functions as secretory organs. They are the **salivary glands, liver**, and **pancreas.**

Mouth

The mouth is the first division of the digestive tube. Its anterior boundary is formed by the **lips,** and posteriorly it leads into the pharynx. Its roof is the **palate** and the floor is formed by the **tongue.** Examine the palate and notice that it consists of two parts: the **hard palate,** located anteriorly, and a posterior one, the **soft palate.** From the free posterior border of the soft palate a small portion hangs down toward the tongue; this is referred to as the **uvula** (L., little grape). Uvula is absent in the cat.

Now examine the dorsal surface of the tongue by pulling it ventrally sufficiently far to tighten it. It may be necessary to cut through the glossopalatine arch on one side. The dorsum of the tongue is covered with numerous elevations or papillae. There are three kinds.

1. The **filiform papillae** are most numerous and are high and narrow. The anterior ones in the cat bear spiny projections with which the animal grooms its fur.
2. Small, rounded **fungiform papillae** are scattered among the filiform and can be seen best along the margins of the tongue.
3. The large **circumvallate papillae** are located near the back of the tongue. In man the vallate papillae are only about 12 or 13 in number and are arranged in the form of a V that lies with its apex directed toward the back of the tongue. In the cat there are four to six. Each papilla is a relatively large, roundish patch surrounded by a circular groove.

There are three types of teeth (see human skeleton):

1. **Incisors** – designed for cutting and located in the front of the jaws.

2. **Canines** – pointed teeth for holding and tearing.
3. **Molars** – adapted for grinding.

There are 32 permanent teeth. In each half jaw from the front to the back are two incisors, one canine, two premolars, and three molars; a total of eight on each side, or an overall total of 16 on each jaw. The cat has three incisors, one canine, three premolars, and one molar in the top jaw and the same for lower jaw, except that there are two premolars instead of three.

The dental formula for man

$$\frac{2-1-2-3}{2-1-2-3}$$

The dental formula for cat

$$\frac{3-1-3-1}{3-1-2-1}$$

Salivary Glands

There are five pairs of salivary glands in the cat. These may be exposed by removal of the skin and the fascia ventral and posterior to the ear.

1. The **parotid** is the largest. It overlies the caudal portion of the masseter muscle. The duct enters the oral cavity adjacent to the upper molar tooth.
2. The **submaxillary gland** lies ventral to the parotid just caudal of the angular process of the mandible. Its surface is smooth and is crossed by the posterior facial vein.
3. The **sublingual gland** is cranial to the submaxillary gland lying between the masseter and the digastric muscles. These three pair of glands are found in man. The cat has two additional pairs.
4. The **molar gland** lies between the muscles and the mucosa of the lower lip.
5. The **infraorbital gland** lies on the floor of the eye orbit. Do not confuse this with the **lacrimal**, or **tear**, **gland**, which lies on the surface of the eyeball near the lateral angle of the eyelids.

Posterior to the mouth is the **pharynx**, a common passageway for air and food. It communicates posteriorly with the esophagus and ventrally with the **larynx**. The **epiglottis** acts to deflect food around or over the glottis, the slitlike opening into the larynx. Through the pharynx food is passed into the esophagus, a soft muscular tube lying below the trachea and extending the entire length of the thoracic cavity, lying dorsal to all the organs within this region. The esophagus passes through the **diaphragm** and joins the cardiac end of the stomach located in this region (the cardiac region). The remainder, and by far the greatest part, of the digestive system is found within the abdominal cavity (Fig. 59).

The stomach is a baglike organ having the shape of the letter J. It has two curvatures, a **lesser curvature** on the right and a **greater curvature** on the left. It is divided into three parts:

1. A central portion called **fundus** (widest portion).
2. A constricted portion leading into the small intestine, the **pylorus.**
3. A **cardiac portion,** which is the opening into the esophagus since it lies under the heart.

The **pyloric orifice** is guarded by a valve, the pyloric valve.

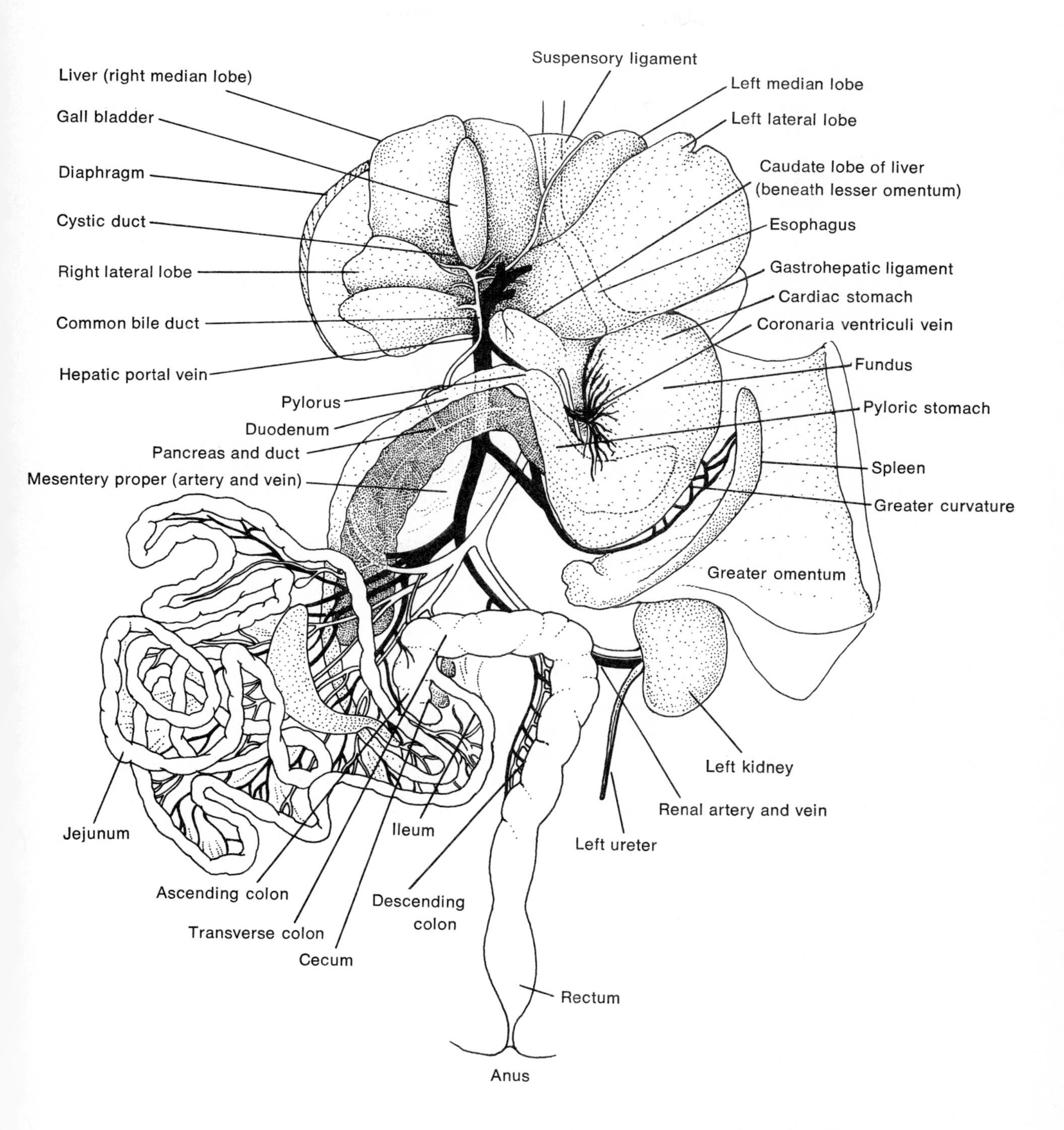

Fig. 59. Digestive tract of the cat showing part of the urogenital system of the female and its relationship to the digestive tract. The intestine, liver, and greater omentum are reflected.

The long, coiled, yellowish tube leading from the pyloric end of the stomach is the small intestine, which extends caudal from the stomach and eventually enters the large intestine. The small intestine is divided into three parts. The first part is referred to as the **duodenum,** the second as the **jejunum,** and the third as the **ileum.** There is no distinct demarcation between these divisions. The small intestine is supported by a thin membrane, the **mesentery.**

The **large intestine** has a larger diameter than the small and is divided into two main portions, the **colon** and the **rectum,** the former making up most of the large intestine. The colon is further divided into three parts: the **ascending colon,** which leads off the small intestine. A short, blind diverticulum, the cecum, extends caudally from the beginning of the colon. The vermiform appendix of man is located at the end of the cecum. The cat does not have one. The ascending colon crosses to the left side to become the **transverse colon,** which extends back into the pelvic canal and becomes the **descending colon** to the rectum and anus.

Liver

The liver, an accessory gland of the gastrointestinal tract, is the largest gland in the body. It lies directly beneath the diaphragm. Notice that more of the liver lies on the right side of the abdominal cavity than on the left. Each half is divided into a lateral and a medial lobe, thus giving a **right** and **left medial lobe** and a **right** and **left lateral lobe.** In the cat the right lateral lobe is split into two **lobules** by a deep cleft. The **gallbladder,** which stores the bile formed by the liver, lies in a depression on the dorsal surface of the right medial lobe. By lifting up the left lateral lobe, you will see a small **caudate lobe** lying deep to a mesentery, the **lesser omentum.**

Pancreas

"Pancreas" comes from the Greek word meaning "all flesh." It is the remaining accessory organ of the digestive tract. It secretes **pancreatic juice** into the small intestine. It lies on the posterior abdominal wall behind the stomach and can be recognized by its lobulated structure. A portion of it lies in the cavity formed by the horseshoe like curve of the duodenum. Another portion stretches across the posterior abdominal wall, the end of which comes in contact with the spleen.

35

Respiratory System

The respiratory system functions to bring air to the inside of the lungs where the oxygen can be extracted from it by the blood. It also functions to rid the body of carbon dioxide. It includes such anatomic divisions as the nasal cavity, the nasopharynx, the larynx, the trachea, the bronchi, and the lungs.

The anterior openings of the nasal cavity are called the **nares,** or **nostrils.** The posterior openings into the nasopharynx are called the **chonae,** or **posterior nares.** The **nasopharynx** is that portion of the pharynx lying directly behind the nasal cavity. In each lateral wall is the opening of the **pharyngotympanic tube** (auditory or eustachian tube), which leads into the cavity of the middle ear.

The **trachea,** or windpipe, is found lying in the median line of the neck. It is ventral to the esophagus and can be recognized by its ringed appearance, caused by **tracheal rings,** or cartilages. At the anterior end of the trachea is the **larynx,** or voice box. The floor of the larynx on the central surface is formed by two unpaired cartilages: (1) the **thyroid cartilage,** a large shield whose caudal margin is indented like an inverted V (forms the Adam's apple) and (2) the **cricoid cartilage,** which is shaped like a signet ring. The cricoid cartilage lies below the thyroid cartilage in front, while posteriorly its wide "signet" part projects upward between the thyroid cartilage. It is the only complete ring of cartilage in the respiratory tract. The two cartilages are joined ventrally by the **cricothyroid ligament** and ventrolaterally by the **cricothyroid muscles.** The inner framework of the larynx consists of the **arytenoid cartilages,** the cartilage of the **epiglottis,** the **quadrangular membrane,** and the **cricovocal membrane.** The **arytenoids** (Gr., like a ladle) rest on the posterior part of the cricoid, one on each side of the midline.

The cartilage of the epiglottis is shaped like a narrow leaf, its rounded end directed upward and its stalls (pointed end) attached in the angle of the thyroid cartilage.

From each side of the cartilage of the epiglottis extends a quadrangular membrane back to the arytenoid.

Extending from the superior edge of the cricoid cartilage into the angle of the thyroid cartilage in front and attaching near the stalk of the epiglottis is the **cricovocal membrane**, or **elastic cone.** Its superior borders are free, and each consists of a well-defined ligament of yellow elastic fiber. These are the two **vocal ligaments.**

Trace the trachea caudally. At the level of the sixth rib, it divides into the two **bronchi**, one for each lung (Fig. 60).

The **lungs** lie in the pleural cavity. The epithelium lining the walls of the pleural cavity is called the **parietal pleura**, and that covering the lung, the **visceral**, or **pulmonary**, pleura. The **right lung** is divided into three **lobes** – **anterior**, **middle**, and **posterior.** The posterior, the largest, is subdivided into a **lateral** and **medial lobule.** The medial is the smallest. The lobes of the lung are attached to the medial wall of the pleural cavity by a pleural fold known as the **pulmonary ligament.** Cut into part of the lung and notice that it is not an empty organ but a very spongy one. The **alveoli** (air sacs) are not visible grossly, but may be seen microscopically (Fig. 61). In the cat the left lung is divided into three lobes. In addition to these three, the right lung has a small cardiac, or azygos, lobe.

The medial wall of each pleural cavity is formed only of a layer of parietal pleura. The potential space between the right and left pleural cavities constitutes the **mediastinum.**

The left lung is divided into two lobes (human).

Examine and carefully note all the structures listed previously. (See Figs. 62, 63, and 64.)

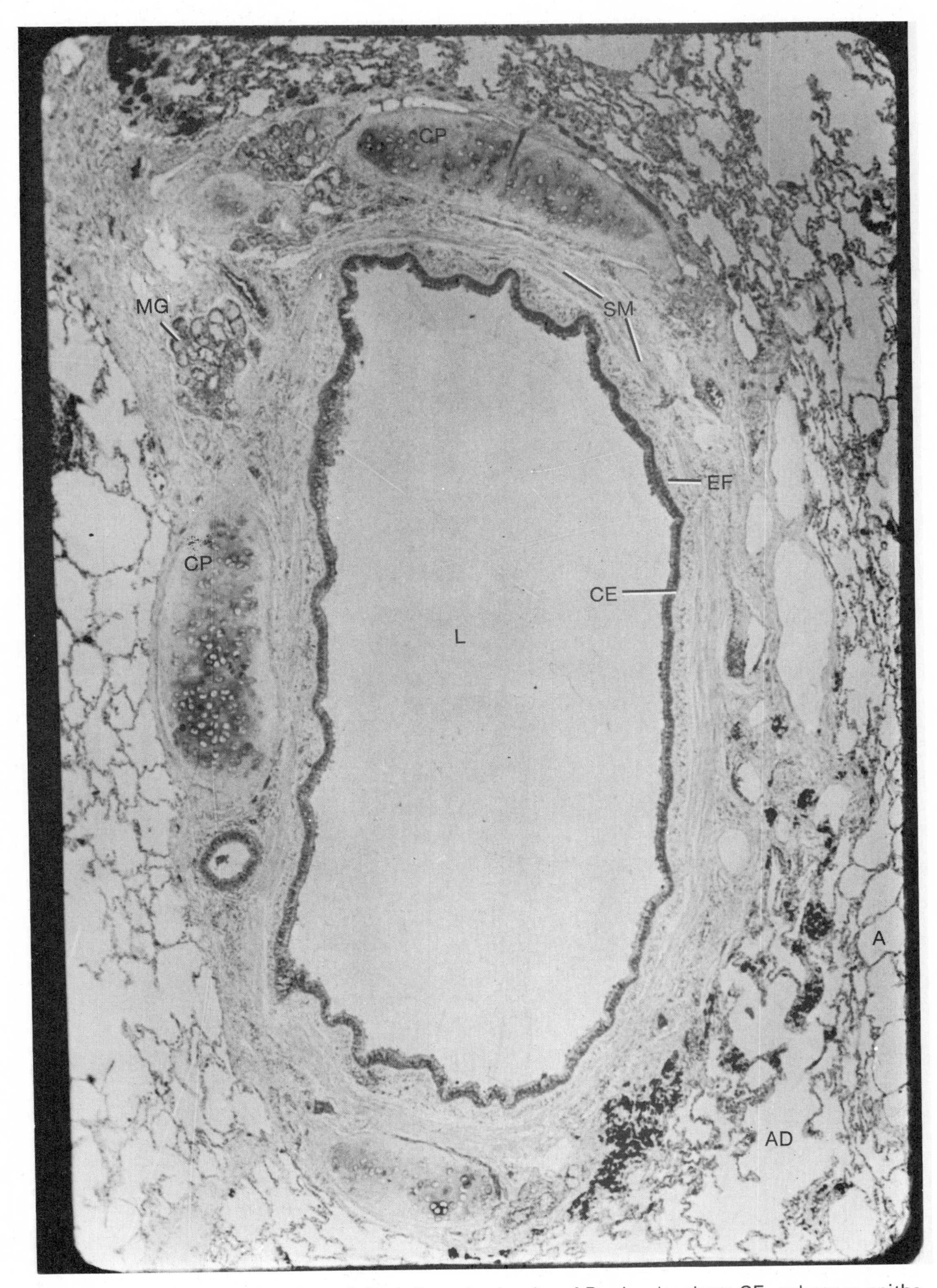

Fig. 60. Lung, cross section through bronchus. A, alveolus; AD, alveolar duct; CE, columnar epithelium; CP, cartilagenous plate; EF, elastic fibers; MG, mucous gland; L, lumen of bronchus; SM, smooth muscle.

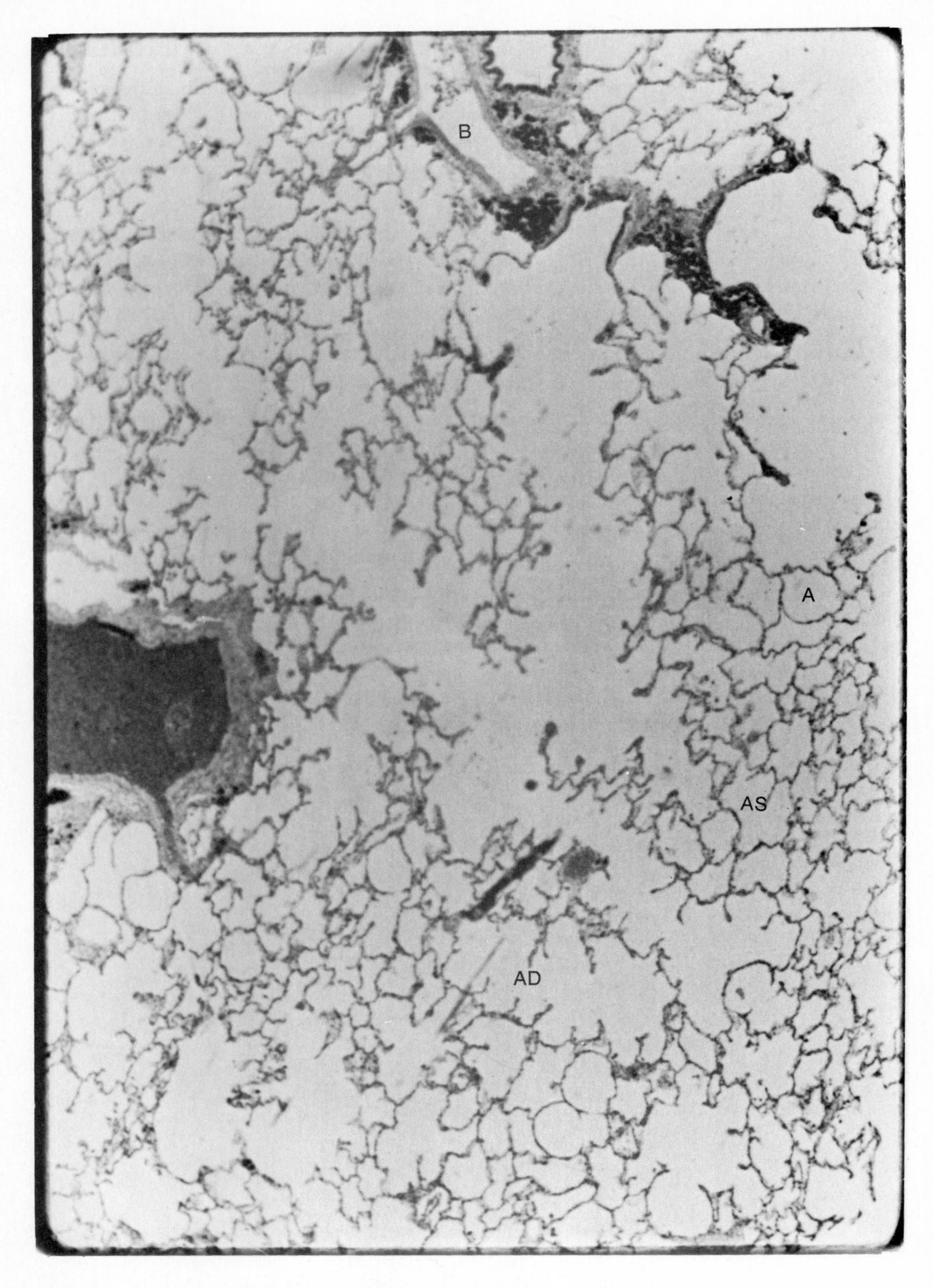

Fig. 61. Lung-respiratory tree showing alveoli. A, alveolus; AD, alveolar duct; AS, alveolar sac; B, bronchiole.

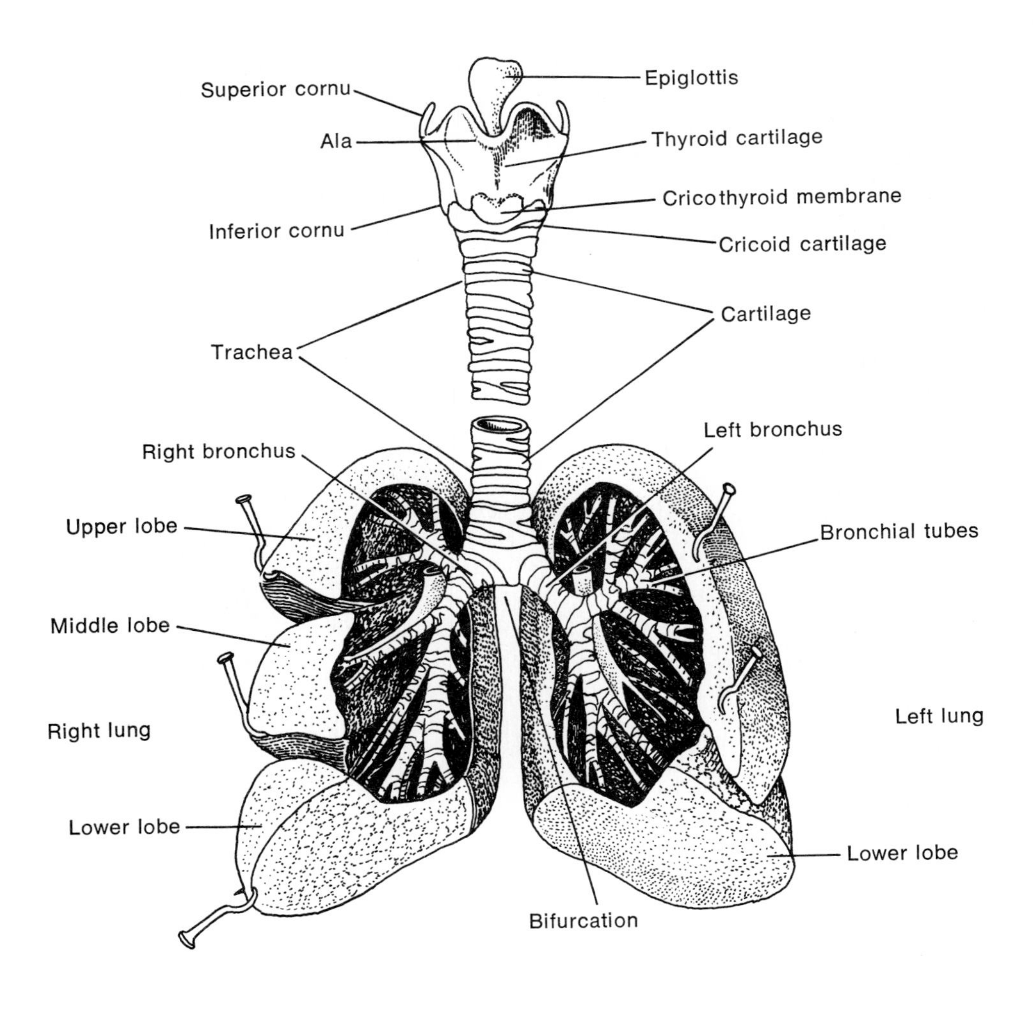

Fig. 62. Front view of the cartilages of the larynx, trachea, and bronchi (human). The lungs have been widely separated and the tissue has been cut away to expose the air tubes.

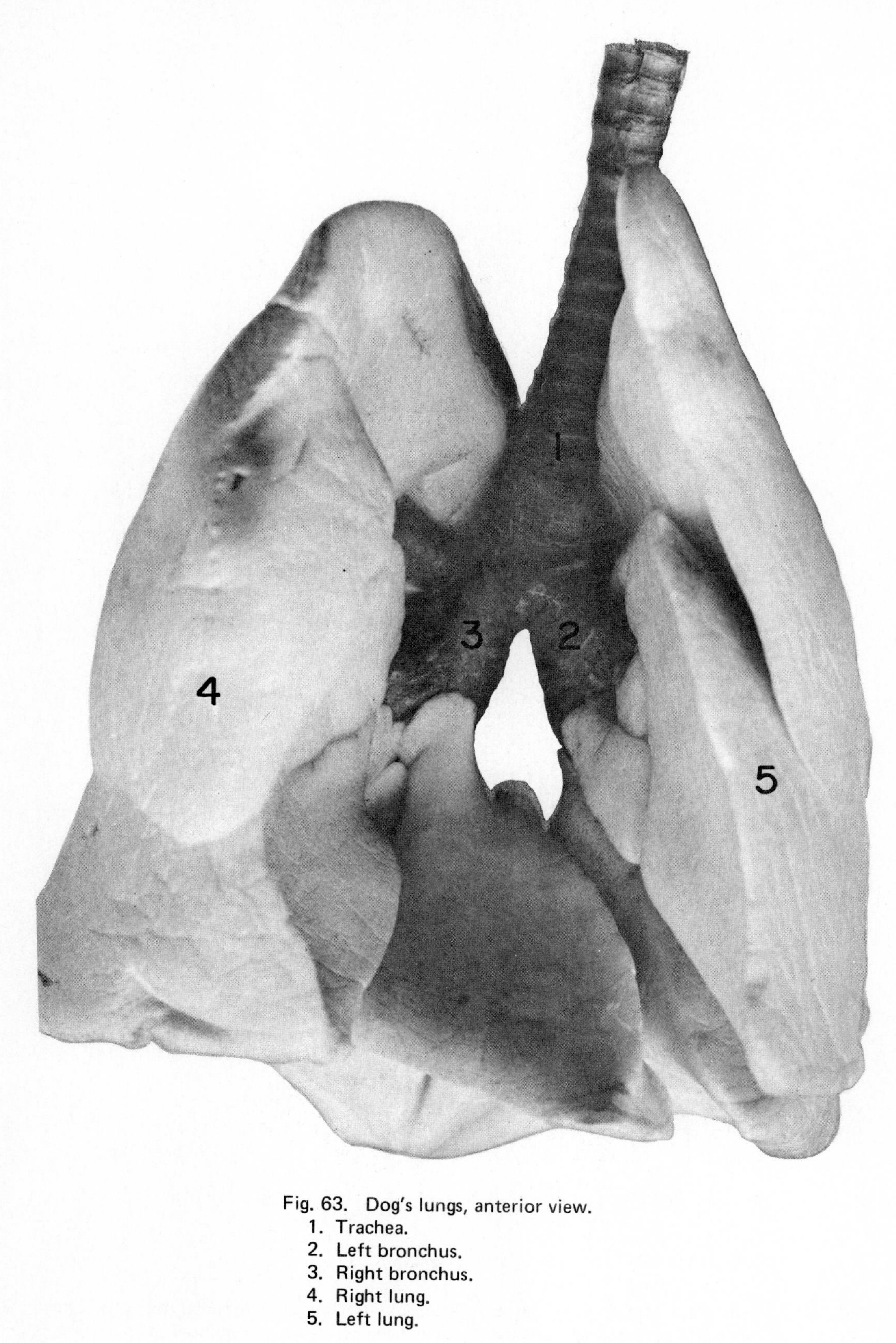

Fig. 63. Dog's lungs, anterior view.
1. Trachea.
2. Left bronchus.
3. Right bronchus.
4. Right lung.
5. Left lung.

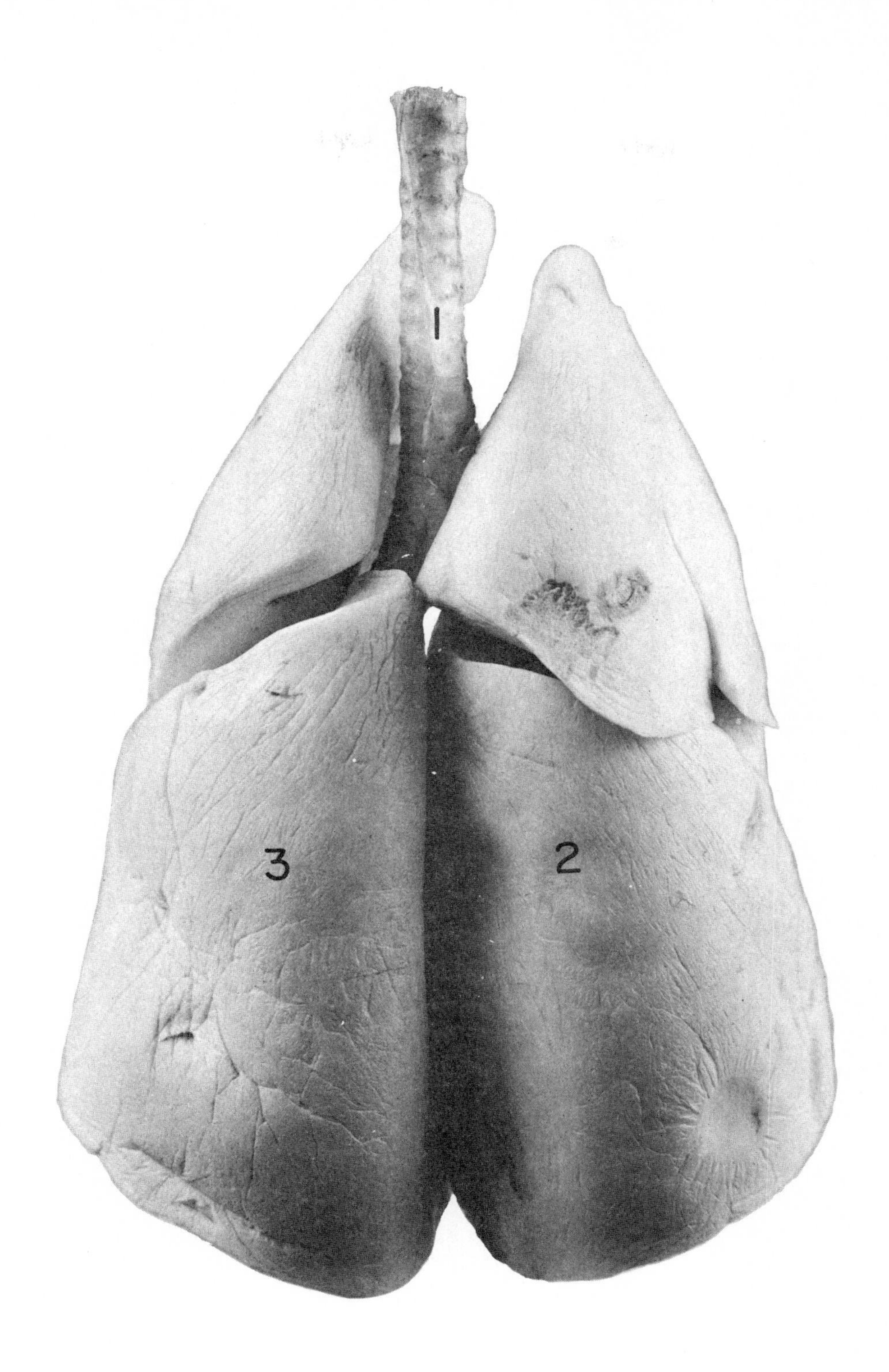

Fig. 64. Dog's lungs, posterior view.
1. Trachea.
2. Right lung.
3. Left lung.

36

Urogenital System

The urinary and genital organs are to be studied at the same time because of their close anatomic relationship.

The **kidneys** are a pair of elongated oval organs with a notch removed from one side so that they actually resemble huge lima beans. In the preserved cat they are usually bluish in color. They lie on the posterior abdominal wall opposite the last thoracic and the first three lumbar vertebrae. The indented portion, directed medially, is called the **hilus,** and here the **ureter** and blood vessels (red – artery; blue – vein) enter and leave the organ. Each kidney is covered with a thin membrane called the **capsule.**

Cut a kidney longitudinally and follow the ureter inward. It opens into a cavity called the **renal pelvis.** A number of large branches called **calyces** are given off, which project still farther into the kidney substance. The solid substances of the kidney can be differentiated into two portions, an outer layer, the **cortex,** and an inner portion, the **medulla.** The medulla is made up of a number of pyramid shaped structures and called **pyramids** of which there are about 12 or 15. Their bases are directed toward the cortex (Fig. 65). A microscopic section of the kidney showing a glomerulus and tubules is seen in Fig. 66.

Find the ureter as it leaves the hilus and follow it as it passes caudad (toward the tail) in a fold of **peritoneum,** which contains fat as it leads into the lower dorsal surface of the **bladder** toward the side.

The **bladder** is a large muscular bag that lies in the pelvis behind the pubic bones. It has three openings: the openings to the **ureters** described above and an opening located anterior to the ureters and more medially for the urethra. The triangular area marked out by the three orifices is called the **trigone.** The urethra passes down and then through the center of the penis in the male. In the female the urethra enters into the lower portion of the vagina.

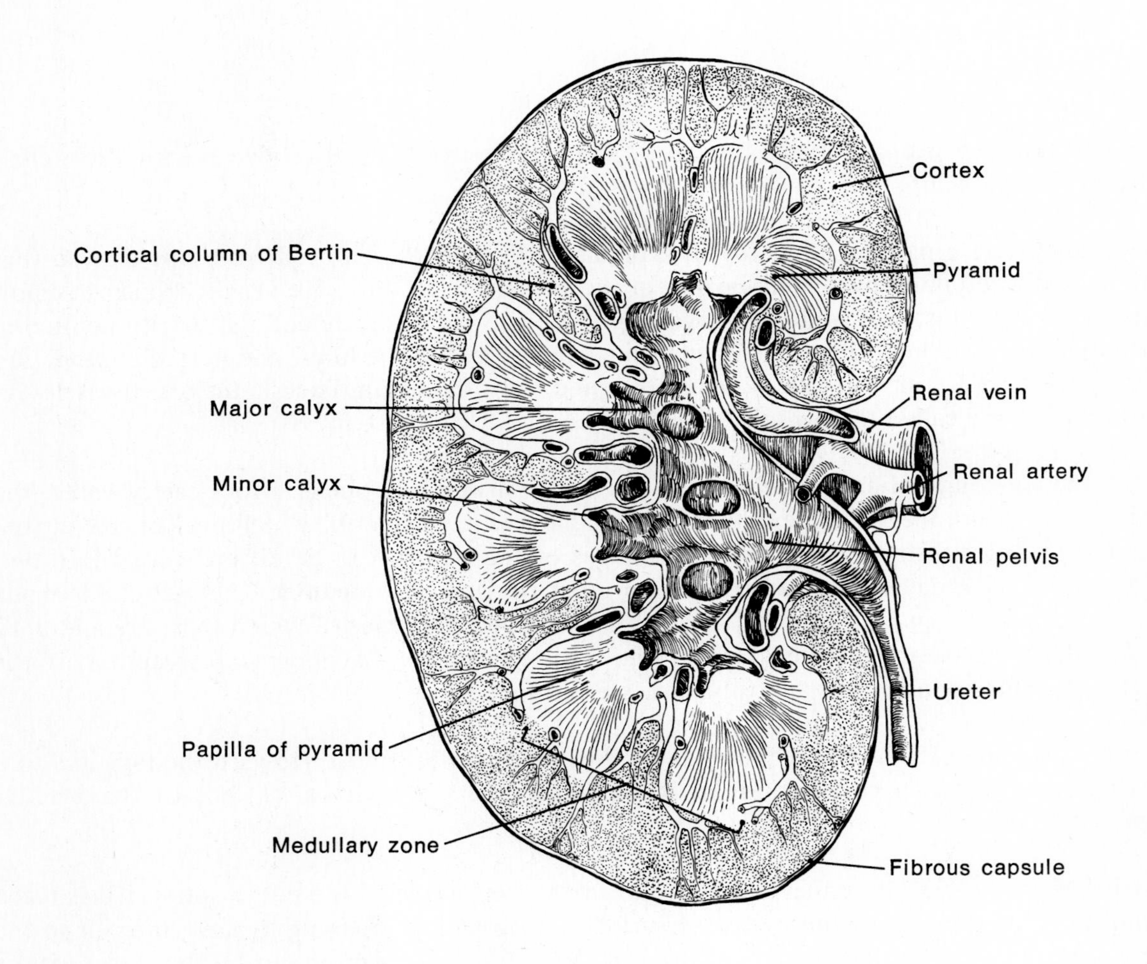

Fig. 65. Longitudinal section of human kidney.

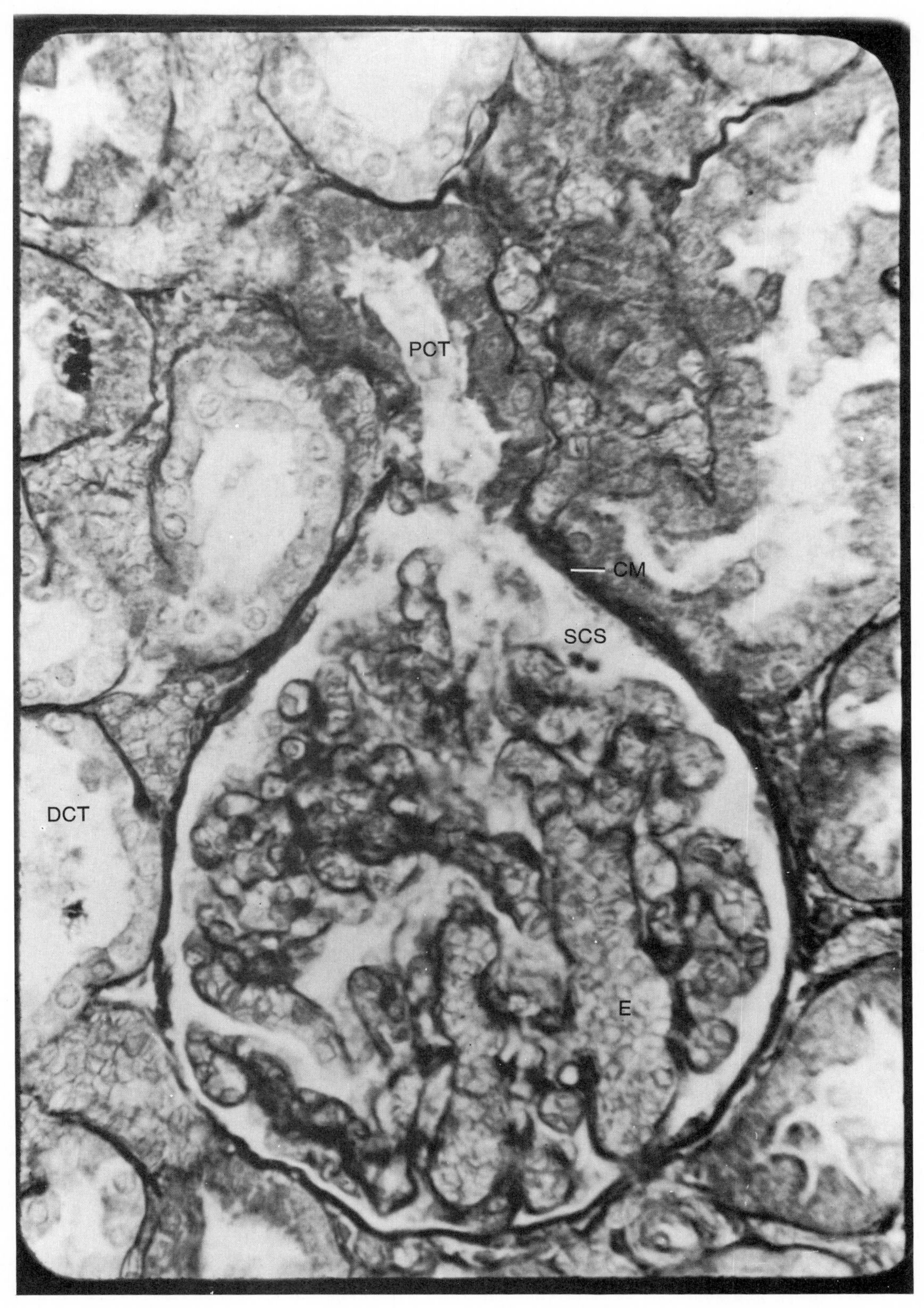

Fig. 66. Cross section of kidney cortex showing glomerulus and tubules; CM, capsular membrane surrounding glomerules; DCT, distal convoluted tubule; E, erythrocytes in glomerulus; PCT, proximal convoluted tubule; SCS, subcapsular space.

Male Genital System (Fig. 67, cat; Fig. 68, human)

The male genital system consists of the following parts:

1. Two glands called the **testes.**
2. A system of ducts (known as the vasa deferentia) that conveys spermatozoa to the urethra; it is enclosed for a short time in the **spermatic cord,** which contains, in addition to the pair of vasa deferentia, a nerve, a vein, and an artery.
3. A **penis,** the copulatory organ, which contains the urethra.
4. A number of accessory glands that secrete fluid – the **prostate,** which surrounds the beginning of the urethra and enters into it, and the **bulbourethral,** or **Cowper's, gland,** which is about 1 in. posterior to the prostate gland.

The testes, which produce the spermatozoa, are located in a baglike structure, the **scrotum,** which is external to the body.

Remove the testes from the scrotum, being careful to keep intact their connections to the spermatic cords that run forward near the median line until they pass through the abdominal wall at the **inguinal canal.** Trace the cord up through this point. It is at this point that the vas deferens leaves the spermatic cord.

It will now be necessary to open the pelvic girdle. With a pair of bone clippers cut through the **symphysis pubis** at the median line. Grasping the thigh of the cat, crack open the girdle still farther. Clip away some of the bone that may still be in the way.

Notice how the vas deferens passes into the triangular enlargement, the prostate gland, and then into the urethra.

Remove the integument from the penis. Note the fold of the integument which forms the **prepuce** or **foreskin,** which covers the enlarged pointed end, the **glans penis.**

Female Genital Organ (Fig. 69, cat; Fig. 70, human)

The female generative organs consist of a pair of **ovaries** and system of tubes, the **uterine** or **fallopian tubes,** the **uterus,** and the **vagina.**

The ovaries are tiny, bean shaped organs lying slightly below the kidneys. The ovaries are attached to the uterine tube by a number of fingerlike processes called the **fimbriae,** which merge into a thin coiled tube – the fallopian tube. These tubes, one connected to each ovary, pass caudad into an enlarged portion of the tube called the horn of the uterus, which then passes into a single tube, the body of the uterus. The overall picture of the tract would then be Y shaped, since there are a pair of ovaries, a pair of fallopian tubes, a pair of uterine tubes known as the **horns of the uterus,** and the single body of the **uterus.** At the lower end of the body of the uterus is the **vagina.** The portion of the vagina in which the urethra enters is referred to as the **urogenital sinus.** The sinus, sometimes called the vestibule, leads to an external opening, the **vulva,** or **pudendal cleft.**

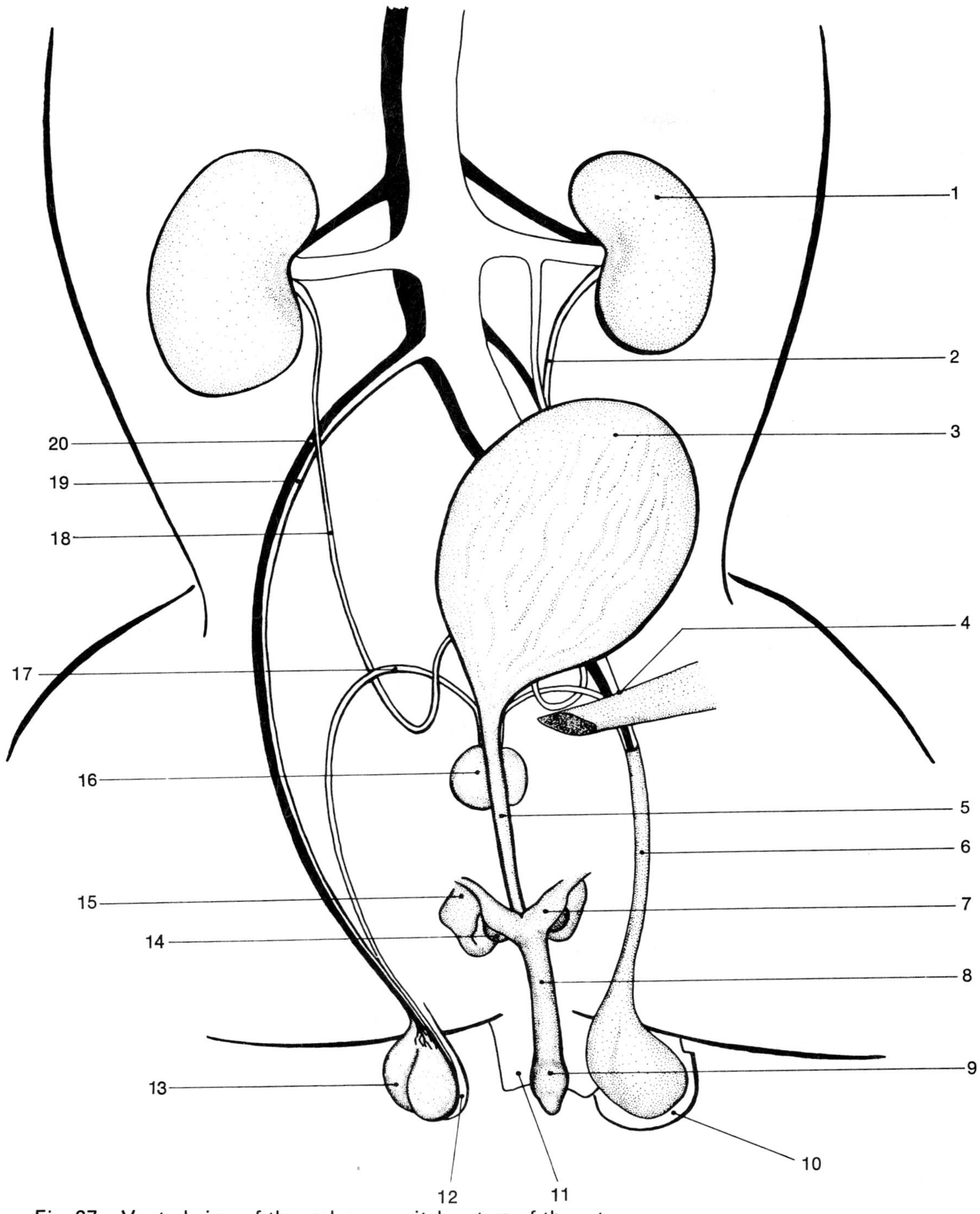

Fig. 67. Ventral view of the male urogenital system of the cat.

1. Kidney.
2. Ureter
3. Urinary bladder.
4. Inguinal canal.
5. Urethra.
6. Spermatic cord.
7. Crus penis.
8. Penis.
9. Glans penis.
10. Scrotum.
11. Prepuce.
12. Tunica vaginalis propria.
13. Testes.
14. Bulbourethral gland.
15. Anal gland.
16. Prostate gland.
17. Ductus deferens.
18. Ureter.
19. Spermatic vein.
20. Spermatic artery.

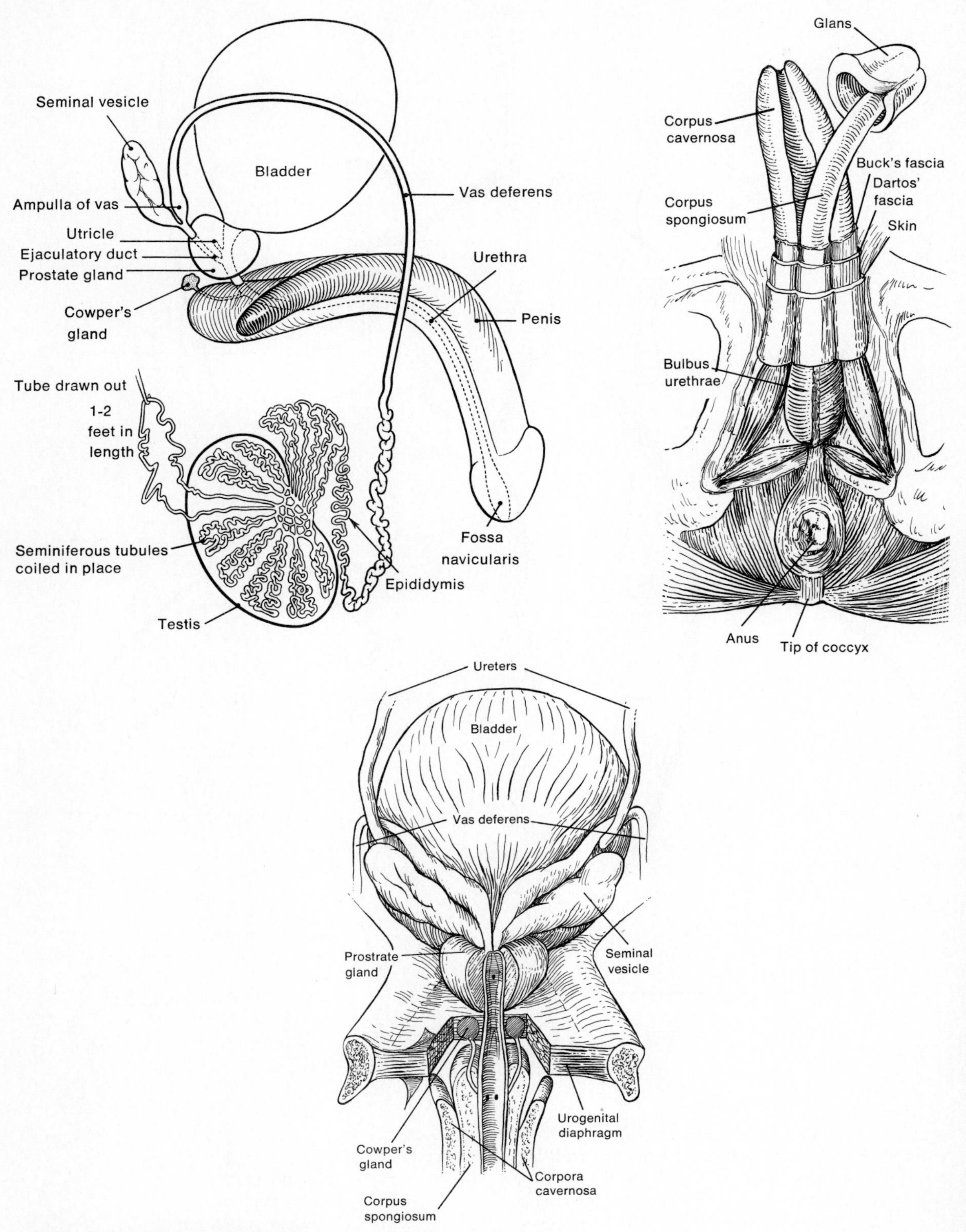

Fig. 68. Male urogenital system of the human.

A. Testis, showing the duct system and the relation of the ducts to the accessory sex glands and penis.

B. Penis, showing its structure.

C. Posterior view of the bladder, showing the relation of the accessory sex glands to the urethra.

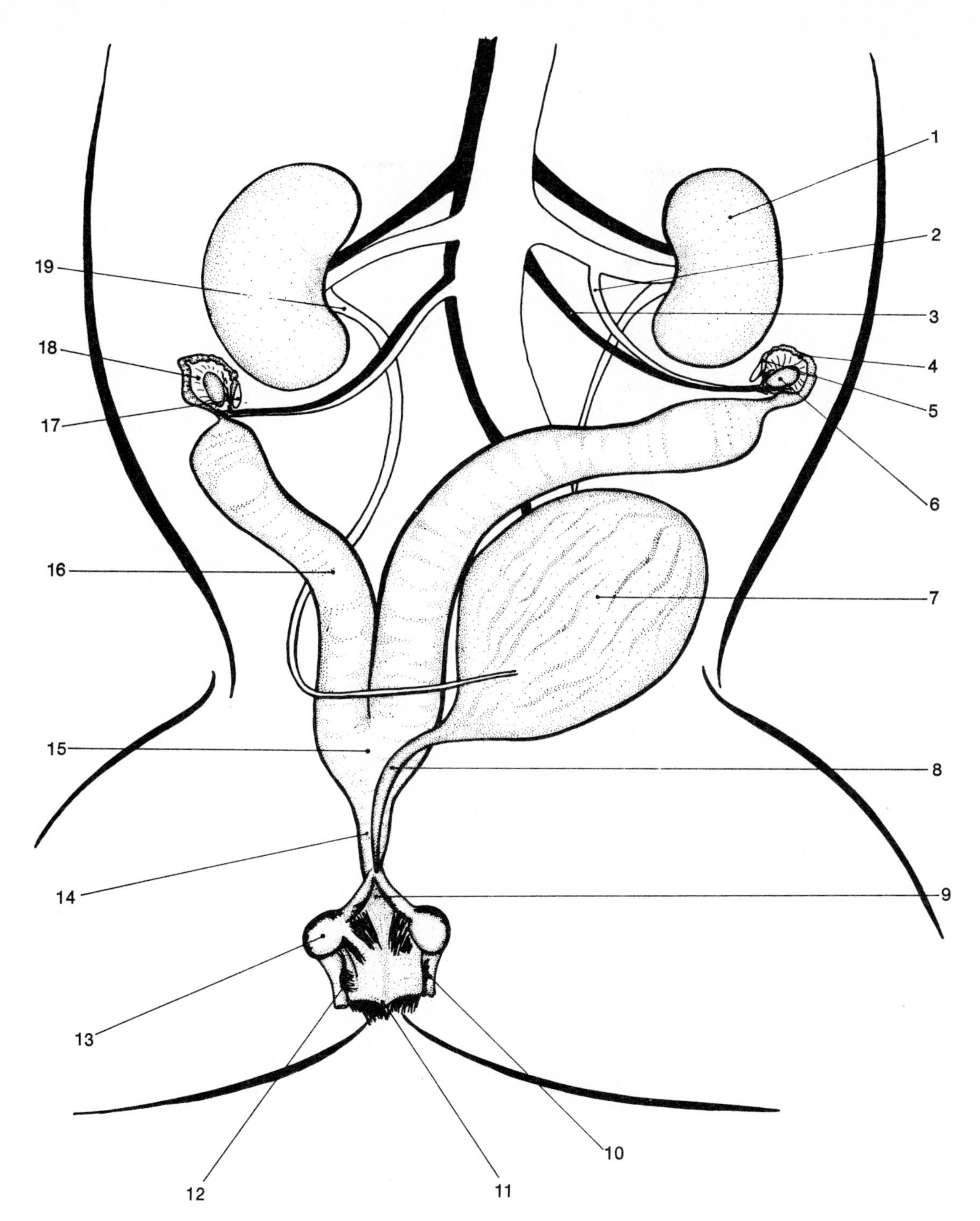

Fig. 69. Ventral view of the female urogenital system of the cat.

1. Kidney.
2. Ovarian vein.
3. Ovarian artery.
4. Uterine tube.
5. Infundibulum.
6. Ovary.
7. Urinary bladder.
8. Urethra.
9. Vestibula.
10. Labium.
11. Urogenital orifice.
12. Clitoris (cut).
13. Anal gland.
14. Vagina.
15. Body of uterus.
16. Horn of uterus.
17. Abdominal ostium.
18. Ovarian ligament.
19. Ureter.

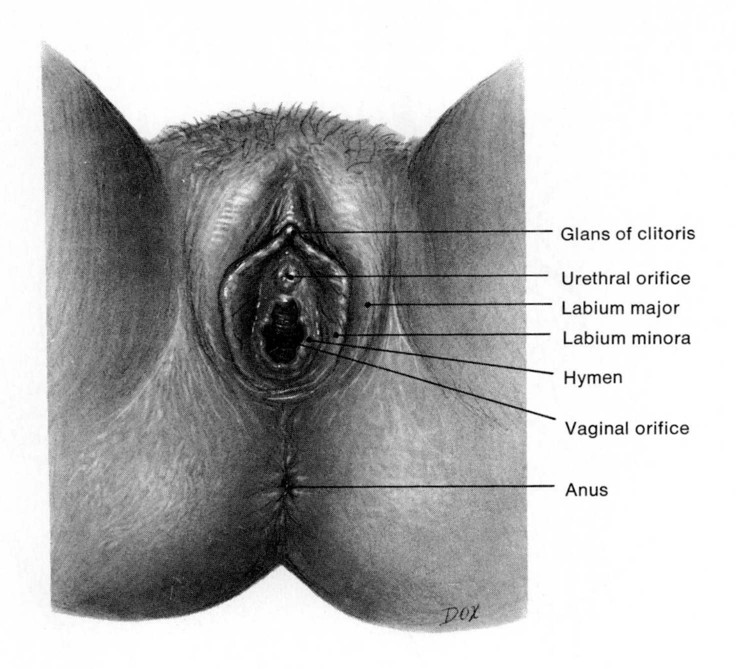

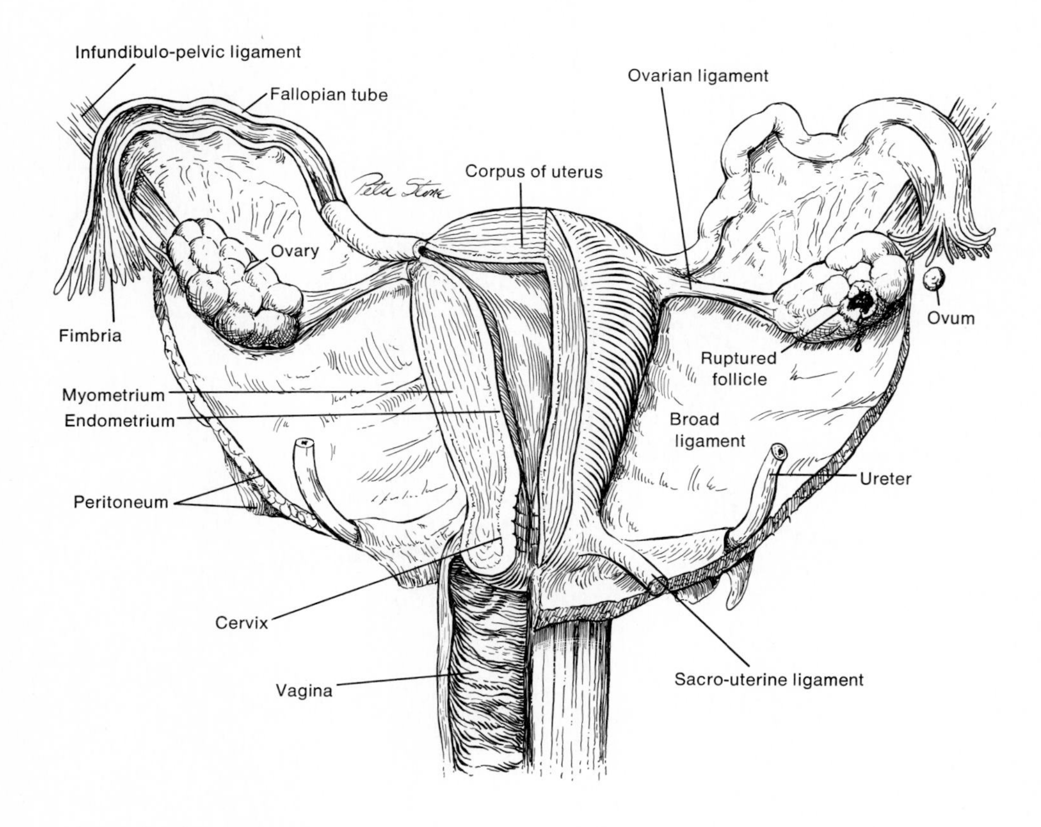

Fig. 70. Female urogenital system of the human.
A. External genital organs.
B. Uterus and associated organs.

In the human female the uterus is of the simplex type and does not divide into right and left horns (bipartite type) as in the cat. The uterine tubes (fallopian tubes) enter the human female's triangular uterus at each angle of its dome shaped base (Fig. 79).

PART II

Physiology

37

Cellular Components of the Blood — Erythrocytes and Leukocytes

The formed elements of the blood include the red blood cells, the white blood cells, and the platelets. Any change in the normal number of these formed elements is important to the clinician, because it indicates that certain physiological changes are occurring within the body. Thus, in many cases, it is important to make a count of the RBC and WBC.

The normal values are

	RBC	WBC
Male	5,000,000/mm^3	5,000–10,000/mm^3
Female	4,500,000/mm^3	5,000–10,000/mm^3

To make a blood cell count, the following materials are needed:

1. Blood lancet
2. Cotton
3. Alcohol
4. Red and white cell diluting pipettes
5. Counting chamber and cover slip (see diagram)
6. Microscope

For all routine blood work, blood is usually taken from the fingertips. One may use the tip of the ear lobe also for some tests. With babies, blood is drawn from the heel or large toe of the foot.

To perform a finger puncture the following are needed:

1. Lancet
2. Finger disinfecting solution (equal parts of acetone and 95 per cent alcohol)
3. Cotton pledgets

PROCEDURE

1. The finger to be punctured is thoroughly cleansed with finger disinfecting solution. (In the meantime the blood lancet should be setting in alcohol for sterilization purposes; it is better to use a disposable lancet.)

2. The point of the lancet is adjusted to type of patient – heavy patient, longer point; light patient, shorter point. The fingertip must be perfectly dry to prevent the blood from flowing off. Discard the first drop of blood unless performing a Tallquist hemoglobin percentage test. You are now ready to prepare the blood for cell counting. Materials needed to make counts are counting chamber, diluting pipettes, and microscope.

3. Red cell count is made in the central primary square (Fig. 71). Always find the central primary square under low power and then switch to high. The central primary square is divided into 25 squares (tertiary squares), and each tertiary square is divided into 16 squares called **quaternary** squares.

4. If the figure directly above the bulb on the diluting pipette is 11, it is a white cell pipette. If the figure above the bulb is 101, mark it as a red cell pipette. Fill pipettes by means of mouth suction and also empty tube by means of mouth. Beads break up clumps of cells and distribute them more evenly through rotations of tubes.

DILUTING FLUIDS

Reasons for use of white cell diluting fluid

Hemolysis (breaking down of hemoglobin or removal of the red cell)
Dilution of blood
Slight staining of the nucleus of white cell
Prevention of bacterial growth and possible formation of molds
Preservations of cell

If in a hurry, make up a 2 to 3 per cent solution of glacial acetic acid and hemolysis will take place – only use in an emergency as a diluting fluid.

White cell diluting fluid

Glacial acetic acid 2, 3, or 5 ml
Distilled water 98, 97, or 95 ml
1 ml of 1 per cent aqueous solution (stain) of crystal violet (gentian violet)
1 g of powdered thymol (preservative) – the diluting fluid will keep without thymol. Thymol is used if blood which has been diluted in the pipette is going to be set aside for future use.

Red cell diluting fluid

In emergencies a 2½ per cent solution of sodium citrate is satisfactory.

Hayem's Fluid

1. Mercuric chloride 0.5 g.
2. Sodium chloride (NaCl) 1.0 g.
3. Sodim sulfate (Na_2SO_4) 5.0 g.
4. Distilled water 200 ml.

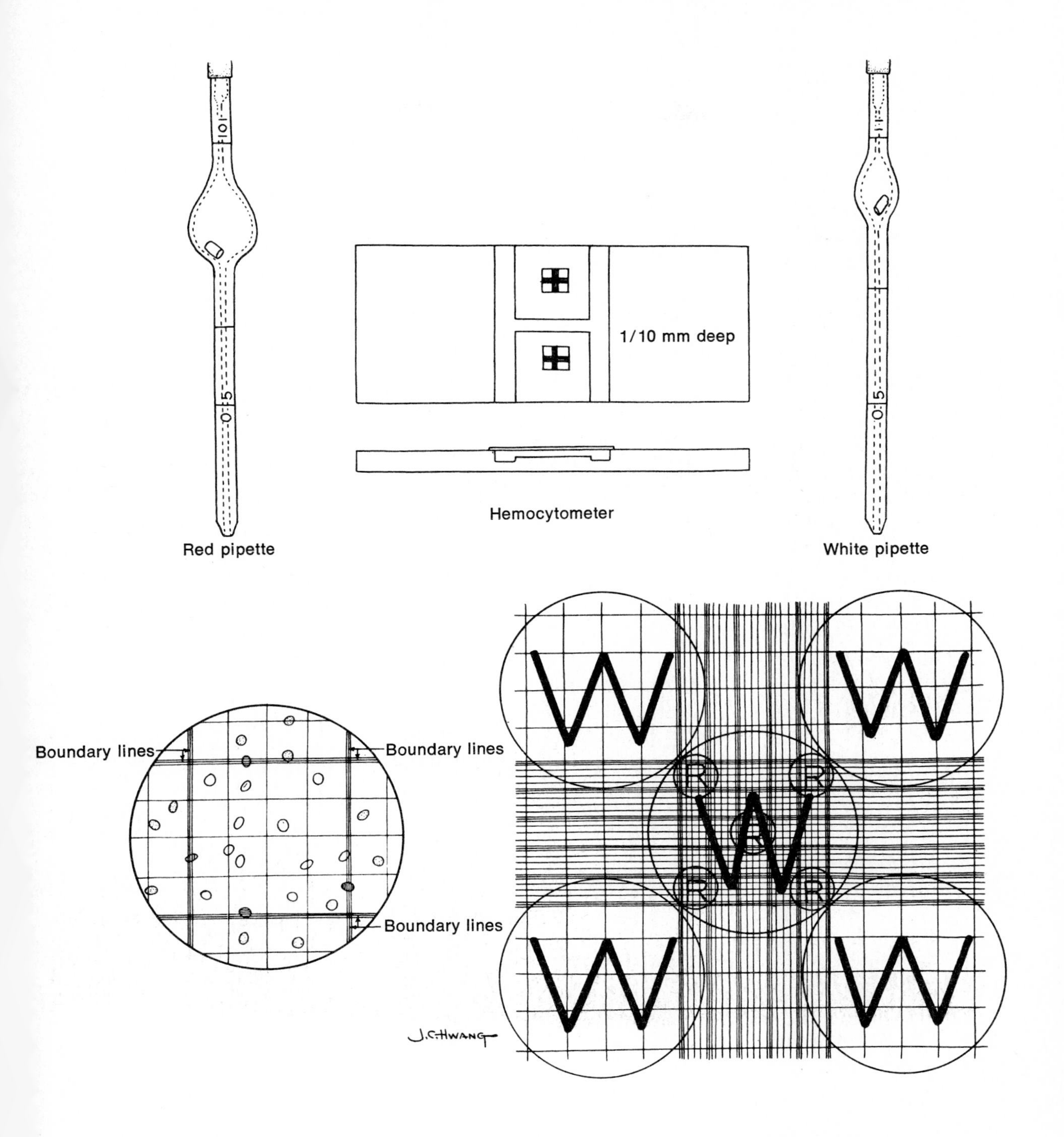

Fig. 71. Blood-counting equipment and magnified areas of the counting portion of the hemocytometer. The red cells are usually counted in the five groups of 16 small squares (R) in the center and four corners of the central square millimeter. To avoid counting a corpuscle twice, those on a line are counted only when they are on the top and left lines, or on the bottom and right lines (see bottom left illustration). The white cells are counted in the four corner 1-mm squares (W). Disregard cells touching two of the boundary lines.

Hayem's fluid made as described, will keep for two weeks.

CLEANING OF HEMOCYTOMETER AND GLASSWARE

Hemocytometer – wash with hot soapy water and rinse; dry with alcohol.

Pipettes

Rinse in cold water.
Rinse in acetone alcohol.
Rinse in ether.

Acetic acid or concentrated nitric acid will break clots.

FILLING PIPETTES

1. Use a clean and perfectly dry pipette; be sure it is the correct pipette.
2. A fresh large drop of blood.
3. Remove cork from diluting fluid.
4. Grasp pipette near point and place on drop of blood and draw up.
5. Keep graduations on pipette facing you.
6. Pass into diluting fluid in horizontal position and draw up; place pipette in vertical position and rotate it so that air bubbles don't form.
7. Remove pipette at a horizontal position; plug one end with finger and remove rubber tubing; plug other end with thumb and shake.
8. Shake pipette for 2 min.
9. Set up hemocytometer; place cover slip on surface; using thumb as control, discard first two drops of blood.
10. Place tip below cover slip at incline of charging surface.
11. If a good charge is received, count ½ min before going into counting of cells.

ERRORS

1. Air bubbles in blood.
2. Pipette pressed in fingers too hard – no blood received.
3. Don't blot end with cotton – plasma is drawn off and increases volume of RBC.

OUTLINE OF RBC

1. Draw blood to 0.5 mark.
2. Diluting fluid 101 mark.
3. Shake for 2 min.
4. Discard first two drops.
5. Charge chamber.
6. Allow cells to settle for ½ min before examination.
7. Check distribution of cells under lower power.
8. Transfer to high dry power – 4 mm (diaphragm always closed in red and white cell counting).
9. Make contact with slip.

10. Look in eyepiece and focus up. Go to every corner of tertiary square and center tertiary square and count all cells in five block square. Add four zeros to total number = total number of cells per cubic millimeter of blood (same as multiplying total count in five tertiary squares by 10,000*).

WBC

1. Blood to 0.5 mark.
2. Diluting fluid to 11 mark.
3. Shake for 2 min.
4. Discard two drops.
5. Charge chamber.
6. Allow to settle for ½ min.
7. Check distribution under low power
8. Count all cells in primary squares W and multiply final result by 50*, which gives you amount of white cells per cubic millimeter of blood.

Count all cells on left and upper lines, but do not count cells that appear on right and lower lines of same square.

List 15 errors that may have been introduced in CBC.

Normals

WBC

Male	5,000–10,000 cells per mm^3 of blood
Female	5,000–10,000 cells per mm^3 of blood

If not normal – recheck

RBC

Male	4,500,000–6,000,000 cells per mm^3 of blood
Female	4,000,000–5,500,000 cells per mm^3 of blood

Difference in number of cells counted in any primary scale should not exceed 10. If it does, make a new charge.

In RBC in central primary scale, the difference between the lowest and highest numbers counted in the designated squares should not exceed 15.

*See next section, "Computing Blood Cell Count Factors."

Computing Blood Cell Count Factors

SUMMARY OF COMPUTING RBC COUNT FACTOR OF 10,000

Red cell count is made in the four corner and center tertiary squares.
Difference in count should not exceed 15.
Count cells on left and upper lines, but not on right lower lines of the same square.
Total the counts in the five tertiary squares.
The five tertiary squares are found in the center primary square.
A primary square has an area of 1 mm^3 (contains 25 tertiary squares).
The count was made in $\frac{5}{25}$ or $\frac{1}{5}$ of the area, or $\frac{1}{5}$ mm^2.
The depth is $\frac{1}{10}$ mm.
The volume is $\frac{1}{50}$ mm^3 ($\frac{1}{5}$-mm^2 area $\times$ $\frac{1}{10}$ -mm depth).
The standard count is for 1 mm^3 (to convert $\frac{1}{50}$ mm^3 to standard, multiply by 50, that is, $50 \times \frac{1}{50} = 1$ mm^3).
The dilution factor is 200 or $\frac{100}{0.5} = 200$
To compare with standard count, multiply total count in the five tertiary squares by 10,000 (50 multiplied by 200).

SUMMARY OF COMPUTING WBC COUNT FACTOR OF 50

White cell count is made in the four W corner primary squares. Do not count in the center W primary square.
Difference in count should not exceed 10.
Count cells on left and upper lines, but not on right and lower lines of the same square.
Total the counts in the four primary squares.
There are 16 secondary squares in each primary square.
A primary square has an area of 1 mm^2.
The count was made in 4 or a total area of 4 mm^2.
The depth is $\frac{1}{10}$ mm.
The volume is 0.4 mm^3 (4-mm^2 area $\times$ $\frac{1}{10}$ -mm depth).
The standard count is for 1 mm^3 (to convert 0.4 mm^3 to 1 mm^3 multiply by 2.5; that is, $0.4 \times 2.5 = 1$ mm^3 .
The dilution factor is 20 ($\frac{11}{0.5}$).
To compare with standard count, multiply total count in the four primary squares by 50 (2.5 multiplied by 20).

Experiment XIX | Blood Tests

BLEEDING TIME

Time required for the cessation of hemorrhage when blood flows through a cut in the skin of measured depth and length (lobe of ear is preferred in this test).

1. As soon as first drop of blood appears, make a note of the time.
2. Blot excess blood off with some sort of blotting paper (do not rub or touch incision or an error will be introduced).
3. Keep blotting excess drops until blood ceases to flow – this is end point of test.
4. Normal bleeding time is from 1 to 3 min (interval between first drop and last drop).

A decrease in blood platelets has an effect in that an increase in bleeding time occurs.

COAGULATION TIME

1. Slide method

   a. Perform finger puncture and place an average size drop of blood on slide.
   b. As soon as blood appears at finger puncture – sight, note time.
   c. At ½ min intervals draw a pin through the blood drop – observe point.
   d. Repeat c until you can pick up fine red threads on end of needle – this is end point. Sometimes entire mass forms a gel, which is also considered end point. Time between first drop to first thread is coagulation time, and normal is from 2 to 8 min.

2. Capillary-tube method

   Capillaries should be 1 to 2 mm in diameter and 4 to 6 in. in length.

   a. Perform finger puncture and note time when blood first appears.
   b. Place capillary in blood drop, so that it becomes practically filled with blood. Do this by holding tube lower than drop.
   c. At ½ min intervals break off small sections of tube until you arrive at one that has fibrin thread between broken parts (this is end point).

An increase in coagulation time may have many causes: for example, diminished prothrombin, calcium, or fibrinogen. Various diseases cause a diminishing of these substances.

Record your bleeding time and coagulation.

	Individual		Lab Section	
	Bleeding	*Coagulation*	*Bleeding*	*Coagulation*
Minimum time				
Maximum time				
Average time				

BLOOD GROUPING

The international classification of blood groups – O, A, B, AB – has been accepted by the National Research Council as the standard method. The letters used indicate the presence of antigens (A, B, AB) or the absence of antigens (O). The following table compares the international system with Moss and Jansky systems.

Comparison of Blood Groups

International	*Jansky*	*Moss*	*% Occurence*
O	I	IV	40–45
A	II	II	40
B	III	III	10–15
AB	IV	I	5

Preparation of sera. At one time commercially prepared dried rabbit sera was standardly used. Rabbits were immunized against known group A and B human sera, and the anti-A and anti-B rabbit sera were dried and stored in sealed containers.

Today human anti-A and anti-B sera are used. It is prepared from humans by collecting blood from young healthy donors of groups A and B. The blood is allowed to clot, and the serum is collected in small containers and sealed. The serum may be stored in these containers at 7°C. It is important that the groups A and B blood be collected separately to avoid error. Serum of blood group A contains anti-B, and serum of blood group B contains anti-A. In order to increase the titer of donor antibodies, A donors are immunized with purified horse B-antigens, and B donors are immunized with purified A-hog antigens.

Determination of blood groups. A glass slide is divided into halves by a wax pencil. The left half is marked "A" and the right half "B". A suspension of red blood cells of the individual to be grouped is prepared by adding one drop of blood to a small test tube containing 2 ml of physiological saline (0.9 per cent saline solution). If rabbit antiserum is used, a drop of the cell suspension is placed on each side of the marked slide. With a toothpick or similar object, a small amount of the powdered anti-A rabbit serum is mixed with the cell suspension on side A.

Similarly a small amount of anti-B rabbit serum is mixed with the cell suspension on the right side marked B. The blood group is determined from the following reactions:

Group O	no agglutination of blood cells on either side of the slide
Group A	agglutination of blood cells by anti-A serum
Group B	agglutination of blood cells by anti-B serum
Group AB	agglutination of blood cells on both sides of the slide

If human grouping sera are used, a drop of A serum is placed on "A" side of the glass slide and a drop of B serum on the left. A drop of the red blood cell suspension to be tested is added to each of the drops of sera. The slide is gently tilted back and forth to cause mixing. The blood group is determined from the following reactions:

Group O	no agglutination of blood cells on either side of the slide
Group A	agglutination of blood cells by B serum (anti-A)
Group B	agglutination of blood cells by A serum (anti-B)
Group AB	agglutination of blood cells on both sides of the glass slide

Commercial sources of human sera

Ortho Pharamaceutical Co., Raritan, New Jersey
Dade Reagents, Inc., Miami, Florida
Michael Reese Research Foundation, Chicago, Illinois

Record your blood group and summarize the results of the entire class in the following table:

Blood Group	*Individual*	*% of Section*
O		
A		
B		
AB		

HEMATOCRIT – RELATIVE PLASMA AND BLOOD CELL VOLUME

Puncture the finger as in the previous experiment and by capillary action fill the small capillary tube (hematocrit tube) with blood. Insert the hematocrit tube in the hematocrit centrifuge and centrifuge at high speed for 4 min.

After centrifugation make the following observations:

What is the color of the plasma?

Where are the blood cells found and why?

Determine the percentage of plasma and of blood cells.

What is the difference between plasma and serum?

Tabulate the hematocrit readings of the class. What are the normal values for hematocrit readings?

Hematocrit Results

Subject	*% Blood Cells*	*% Plasma*
Average		

Hemoglobin—Estimation of Percentage in Grams per 100 m

MATERIALS AND EQUIPMENT

Lancet, alcohol, 70 per cent, cotton, distilled water, hydrochloric acid 0.1 N – (30 ml), Tallquist booklet, Sahli outfit

PROCEDURE

There are many methods of estimating the amount of hemoglobin (Hb) in the blood; several are used in this exercise. Different workers have made separate investigations of the percentages of hemoglobin normally present, and, depending upon in what part of the world this work was done, the standards they used vary. Thus, in the Midwest the average normal amount of hemoglobin in grams per 100 ml of blood may vary considerably from the normal average in some parts of Europe or Asia. There are many factors that enter into the reason for these different normals, such as altitude, mineral content of the water, diet, and so on. The important thing to remember is that, when percentage values of hemoglobin are discussed, the standard value (grams Hb per 100 ml of blood taken as the normal average for the particular testing method) should always be mentioned.

A. The Tallquist Method. The student will be supplied with a Tallquist booklet containing blotting papers and a number of strips colored in different intensities of red (Tallquist hemoglobin scale) for comparison. A drop of blood is drawn upon a piece of the blotting paper and the color obtained is then compared with a standard hemoglobin scale. The number opposite the strip that matches your blood in color is the per cent Hb in your blood. The Tallquist standard is usually 15.6 g 100 ml of blood, which is considered 100 per cent hemoglobin. The per cent figure indicated by comparing the blood stain with the standard should be multiplied by 15.6 (or by the given standard if different) and the product divided by 100 in order to ascertain the grams Hb per 100 ml of blood.

Example:

per cent reading from Tallquist Standard Chart = 90 per cent

$$\frac{90 \times 15.6}{100} = \frac{1404}{100} = 14.04 \text{ g 100 ml of blood}$$

B. The Sahli Method. 0.1 *N* hydrochloric acid is placed in the graduated tube of the Sahli outfit so that the bottom of the meniscus stands at the mark 10 on the scale. Draw blood into the capillary pipette to the mark 20. Transfer the blood into the acid solution in the graduate. Be sure the point of the pipette is beneath the surface. Rinse the pipette 2 or 3 times with the contents of the graduated tube. Shake the tube and allow it to stand 1 min. Add distilled water a drop at a time, mixing after the addition of each drop, until the color of the tube corresponds exactly to the color of the standard. Read directly from the graduated tube the per-cent Hb and grams of Hb per 100 ml of blood.

What is the standard for the Sahli apparatus you have? (Generally it is 17.3 g of hemoglobin per 100 ml. Thus 17.3 g is considered 100 per cent)

Tabulate your results from both methods in the following table:

Female Subjects	*Tallquist*	*Sahli*	*Male Subjects*	*Tallquist*	*Sahli*

Specific Gravity of Blood

The specific gravity of the blood may vary under certain conditions. Among other things, it gives us information regarding the water content of the blood. A specific gravity greater than normal would suggest that there is less water in the blood than normal and vice versa. There are several methods of determining the specific gravity, two of which are given.

MATERIALS AND EQUIPMENT

Frog (1 for 3 or 4 groups) or rat, lancet, alcohol, 70 per cent, cotton, test tubes–1 X 15 cm (5), chloroform (100 ml), benzene (100 ml), Hydrometer–specific gravity 1,060, graduated cylinder, 100 ml

PROCEDURE

Hammerschlag's Method. Fill a hydrometer cylinder about two thirds full of a mixture of benzene and chloroform having a specific gravity of 1.059. Puncture the tip of the finger with a sterilized lancet and allow a drop of blood to fall into the mixture. If the drop of blood floats on the surface, it is lighter than the mixture. In this case, benzol is added drop by drop, with constant stirring, until the drop of blood neither rises to the surface nor sinks to the bottom. If, when the drop of blood is first drawn, it sinks to the bottom, its specific gravity is greater than that of the mixture, and chloroform is added until the blood remains suspended. The specific gravity of the mixture is then determined by means of the hydrometer.

Copper Sulfate Solution Method. An alternate method of determining the specific gravity of the blood is by means of a series of copper sulfate solutions. Prepare five (or more, if desirable) solutions of copper sulfate whose specific gravities range from 1.050 to approximately 1.064. Be certain that, between the extremes of these solutions, one will have the specific gravity of 1.059. *The correct quantities of copper sulfate for such solutions can be obtained from any recent handbook of physics and chemistry.* To determine the specific gravity of the blood, proceed as under Hammerschlag's Method, but instead of adjusting the specific gravity of any solution, drop blood into successive solutions until the blood remains suspended in one. The specific gravity of this solution will be the same as that of the blood.

Test Conditions. Using either of these two methods, determine the specific gravity of the blood under the following conditions:

1. Have the subject remain at rest for at least 30 min.
2. Let the subject exercise strenuously until out of breath. Test a sample of blood immediately. What is the specific gravity? Explain why exercise has changed the specific gravity of the blood.
3. Determine the specific gravity of the blood 1/2 hr and 1 hr after drinking at least a pint of water.

Can one draw any conclusions as to the extent and rapidity of water absorption?

4. What is the class average for the normal specific gravity?

Tests for Blood in Body Fluids or Tissues

MATERIALS AND EQUIPMENT

Lancet, alcohol, 95 per cent, cotton, guaiac solution (1 g of guaiac resin in 60 ml of 95 per cent alcohol), hydrogen peroxide, benzidine-glacial acetic acid solution.

PROCEDURE

The Guaiacum Test for Blood. Many tissues and body fluids including blood contain oxidizing enzymes. These enzymes are capable of activating hydrogen peroxide so that activated oxygen is produced, which is then available to oxidizable substances in the cell or surrounding fluid. If **tincture of guaiacum** (containing guaiaconic acid) is exposed to "active" oxygen, it turns blue. If a few drops of tincture of guaiacum are added to tissue extracts along with a few drops of hydrogen peroxide, it turns blue. The hydrogen peroxide is activated by an enzyme (catalase) in the extract and the oxygen released is utilized.

1. To a small amount of blood (1 drop in 20 ml of distilled water) add a drop or two of guaiac solution and a small amount of hydrogen peroxide.

What happens?

2. To a small piece of frog muscle, washed free of blood and minced to break up the tissues, add a drop of the tincture and hydrogen peroxide.

What is the result?

3. Test a drop of saliva the same way. What is the result?

Test for Blood in Urine. Place 5 ml of urine in a test tube by means of a pipette. Add some guaiac solution into the urine until a turbidity results. Add hydrogen peroxide, drop by drop, until a blue color is obtained.

Could this be used as a specific test for blood? Explain briefly.

Benzidine Test for Blood. The benzidine method is one of the most sensitive provided the reagents are of satisfactory quality.

The following reagents are used:

Benzidine – a saturated solution in glacial acetic acid. Store in a dark brown bottle – keeps well for some time.
Hydrogen peroxide – usual 3 per cent solution.

Proceed as follows: Dilute a drop of mammalian blood with 20 ml of distilled water. Place 1 ml of this mixture in a clean test tube. Do the same with a drop of frog's blood. Use 1 ml of water in another test tube for control. To each test tube add a few drops of benzidine-glacial acetic acid solution and a few drops of hydrogen peroxide.

Result: If blood is present a green to deep blue color develops, the depth of color depending on the amount.

Blood in feces may be determined using the benzidine test by smearing a sample of feces on a glass slide or filter paper and pouring the reagents over it. The intensity of blue color and the speed with which it develops are a rough measure of the amount of blood present.

Experiment XX | Automaticity of Heart — Rate of Beat

MATERIALS AND EQUIPMENT

Frog, frog board, silk thread, 10 per cent urethane solution, hypodermic syringe, penhook (S shaped), metal rod, kymograph, signal magnet, heart lever.

PROCEDURE

1. Anesthetize frog by injecting 1 ml of urethane solution into dorsal lymph sac.

2. Expose heart by removing skin in thorax region and then by a longitudinal cut through the body wall. Keeping a little to the right of the midline, cut through shoulder girdle. Pull the forelegs well apart and clip back on frog board.

Observe the following:

a. Pinecardium – pericardium
b. Two atria
c. Ventricle
d. Truncus arteriosus – springs from the ventral surface of the ventricle and divides into two aortae
e. Sinus venosus on dorsal surface of heart: appears as a white fine crescent shaped structure – pick apex of heart up and lift forward to see sinus venosus
f. Right and left superior venae cavae and the inferior vena cava

Make a diagrammatic sketch of the heart.

Now continue the experiment.

3. Keep the heart moist with Ringer's solution.

5. Pass a fine pen hook in the tip of the heart, that is, its apex, and arrange the apparatus (kymograph, smoked drum, heart lever, and signal magnet) so that the heart beat is recorded on the drum. This may be accomplished by tying a piece of thread to the pin and attaching the other end to the heart lever. The lever will move up upon contraction of the ventricles and return to the base line on relaxing. Make a time mark every 30 sec with the signal magnet on the record paper just below the tracing of the heart beat.

6. Having obtained a record of the normal heart beat, touch the upper part of the right atrium (region of the SA node) with the metal rod, which has been heated. Note the increase in rate. Let the beat return to normal and, having cooled the rod in ice water, repeat. The rate decreases. Record beats for 30 sec for each condition. Fill in chart:

	First Trial (beats/minute)	*Second Trial (beats/minute)*
Room temperature		
Cold rod		
Warm rod		

7. Now cut the heart out of the thoracic cavity and place in a watch glass containing Ringer's solution. Count the rate after several minutes. Is the heart beat dependent on the central nervous system? What effect does the nervous system have on the heart?

Experiment XVIII

Cardiac Cycle and the Effect of Various Salts and Drugs on Heart Action

MATERIALS

Same as the previous experiment. In addition, the following solutions:

Amphibian Ringer's with excess calcium ions (99 ml of Ringer's with 1 ml of 10 per cent $CaCl_2$)
Ringer's with excess potassium ions (99 ml of Ringer's with 1 ml of 10 per cent KCl)
Adrenalin
Acetylcholine (1/10,000 conc.)

PROCEDURE

1. Prepare frog as in previous experiment and attach heart to heart lever so that beats can be recorded. The heart must be raised from the body of the frog so that the contraction of each cavity – atria and ventricles – will move the lever. The signal magnet pen should be in alignment with that of the heart pen and should be marked off in seconds. The drum should be running at about 1 cm/sec. Record about 15 heart beats and label the tracing to indicate the peak made by the contraction of the atria and contraction of the ventricle. Approximate the time it took for atrial diastole and systole. Ventricular diastole and systole. Of how many phases is the cardiac cycle composed?

2. Inject directly into the ventricle 1 ml of calcium ion solution.

   Record results.

3. After heart returns to normal, inject 1 ml of potassium ion solution.

   Make record.

4. After recovery, inject 0.5 ml of a 1/10,000 concentration of adrenalin.

   Record.

5. Repeat, after recovery, with 0.5 ml of a 1/10,000 acetylcholine solution.

   Record.

Summarize your results in table form, indicating the effect of various solutions and drugs on heart rate, systole, and pressure.

Experiment XXII Human Blood Pressure Determinations by Sphygmomanometers

It should be clearly recognized that arterial pressures cannot be measured with precision by means of sphygmomanometers. Direct registration of pressures of calibrated intraarterial manometers has shown (1) that even during quiet breathing and slight sinus arrhythmia, systolic and diastolic pressures vary from beat to beat by several millimeters of Hg, and that these differences are greatly intensified during states of arrhythmia and deep breathing; (2) that auscultatory systolic readings from the brachial artery average 3 or 4 mm Hg too low and show average scatter of ± 8 mm Hg; and (3) that auscultatory diastolic pressures taken at the point of dulling of the sounds average about 8 mm Hg too high. The errors of clinical measurement of blood pressure can be summarized by saying that in normal persons a mean error of ±8 mm Hg may be expected in individual readings of systolic and diastolic pressures.

APPARATUS

A sphygmomanometer consists of (1) a compression bag surrounded by an unyielding cuff for application of an extraarterial pressure; (2) a manometer by which the applied pressure is read, (3) an inflating bulb, pump, or other device by which pressure is created in the system, and (4) a variable, controllable exhaust by which the system can be deflated either gradually or rapidly.

TECHNIQUE

1. **The patient.** The patient may be either in a recumbent position or comfortably seated. The patient should be placed at ease and time should be allowed for recovery from any unusual recent exercise, meals, or apprehension. The arm should be bared, slightly flexed, abducted, and perfectly relaxed. In the sitting position the forearm should be supported at heart level on a smooth surface. The hand may be pronated or supinated later, depending on which position is found to yield the clearest sounds. The deflated bag and cuff should be applied evenly and snugly around the arm with the lower edge about 1 in. above the antecubital space. If the veins of the forearm are prominently filled or there is evidence of congestion, the cuff should be applied while the arm is elevated in order to promote venous drainage.

2. **General precautions.** The mercury column must be vertical. The meniscus should be read at a level with the observer's eye. It is not important to place the manometer at the heart level. The sounds heard on auscultation are not heart sounds; some misconception regarding this still exists.

3. **Determination of systolic pressure by the auscultatory method.** A stethoscope receiver should be applied snugly over the artery in the antecubital space, free from contact with the cuff. The pressure in the sphygmomanometer should then be raised rapidly and decreased slowly, as in the palpatory method, until a sound is heard with each heart beat. Note the reading as systolic pressure.

As a rule, systolic pressure determined by the auscultatory method is slightly higher than the pressure at which radial pulse beats are first palpable. In case the palpatory reading should be higher than the auscultatory, a number of maneuvers may be undertaken to improve conditions for hearing the sounds. If, despite such efforts, the pressure read by palpation continues to be higher, it should be accepted as the reading for systolic pressure. This should be noted.

4. **Determination of diastolic pressure by the auscultatory method.** With continued deflation of the system below systolic pressure at a rate of 2 to 3 mm Hg per heart beat, the sounds undergo changes in intensity and quality. As the cuff pressure approaches the diastolic, the sounds often become dull and muffled quite suddenly and finally cease. It appears that the point of complete cessation is the best index of diastolic pressure. Under hemodynamic conditions, in which no cessation of sounds occurs, the point of muffling should be taken as diastolic pressure, if distinctly heard, and should be recorded as the point of muffled sounds. When no clear demarcation of the muffling is heard, diastolic pressure should be left indefinite and so indicated, for example, "150/30?".

5. **The operator.** A number of consecutive determinations of systolic and diastolic pressure should be made, and, if they agree reasonably well, the average should be taken. Variations in auditory acuity are important. A physician aware of defective hearing should use an amplifying stethoscope. Auditory acuity is of importance in subtler ways. The first sounds that occur at systolic cuff pressures and the last ones to occur in determinations of diastolic pressure are exceedingly feeble and of short duration. Physicians improve their acuity by training, but personnel to whom blood pressure determinations are often relegated do not have the equivalent auscultatory experience.

CLINICAL TESTS

Several tests to determine the variability of the blood pressure are used to study the autonomic nervous system in relation to the hypertensive state.

Cold pressor test

1. The blood pressure is determined from the right arm with the subject sitting.

2. The left hand is immersed in ice water for 60 sec and a pressure reading is taken. Pressure readings are also taken in 30 and 60 sec after removal of the hand from ice water, then every 2 min until the readings return to normal.

3. Record your results and plot a graph using time as the abscissa and blood pressure as ordinate. Determine average rise of systolic and diastolic pressures of your section. (Class results should be recorded on the blackboard.)

Breath-holding test

1. Determine blood pressure from the right arm with the subject sitting.

2. At the end of a normal expiration the subject compresses his nostrils with the left hand and closes the lips tightly. In this manner the breath is held for 20 sec and the pressure is taken.

3. Continue to take pressure readings every 30 sec for 2 min after each breath-holding test.

4. Plot your results on graph paper using blood pressure as ordinates and time as abscissa. An increase of 40 mm Hg or more is considered to be significant. Normal systolic and diastolic pressure seldom exceeds 20 mm Hg.

Experiment XXIII | Electrocardiograph

MATERIALS AND EQUIPMENT

A self contained electrocardiograph (ECG) and limb electrodes, electrode jelly, or, a set of ECG electrode plates and cables, cardiac preamplifier, output cable, recorder, and chart mover.

In all living tissues and organs we find evidence of some sort of an electrical field, which may be modified suddenly during activity. The heart is no exception. In fact, it is one of the body organs that best displays bioelectrical activity. With each beat, the cardiac muscle generates an electrical field strong enough to be led off through electrodes held in the hands or attached to the feet. That is, these currents spread out over the whole body and, therefore, may be detected in this way. The electrical waves are not of any considerable magnitude. The voltage at the heart itself is small, and when led off through electrodes held in the hand equals only 2 or 3 mV. Such low voltages require amplification to record them. Usually a high gain, low noise, ac/dc differential amplifier is interposed between the electrodes and electronic recording pen.

Electrocardiograms are permanent records of these electrical variations, which accompany or, probably more correctly, precede the heart beat. They are records of the events of each cardiac cycle in a series of waves or deflections of a coil recording pen, which is a very large galvanometer with enough energy to drive an ink writing stylus. These deflections are designated by the arbitrary letters P, Q, R, S, and T (Fig. 72).

Study has shown that the P wave alone is associated with atrial activity. It begins a brief fraction of a second after contraction starts and ends before the atrial beat ceases. It indicates that during their contraction the atria become electronegative to the ventricles. The other positive and negative waves are related to ventricular activity. The first one is the QRS wave. The R spike is the most prominent part of the wave. The QRS appears a brief fraction of a second before systole of the ventricles and subsides as the pressure within them increases. It is evidently the electrical sign of the excitation process that throws the ventricular muscle into contraction. The T wave is evidently related to the ejection phase. It is variable in shape, is sometimes inverted, and is used extensively for detection of certain heart diseases.

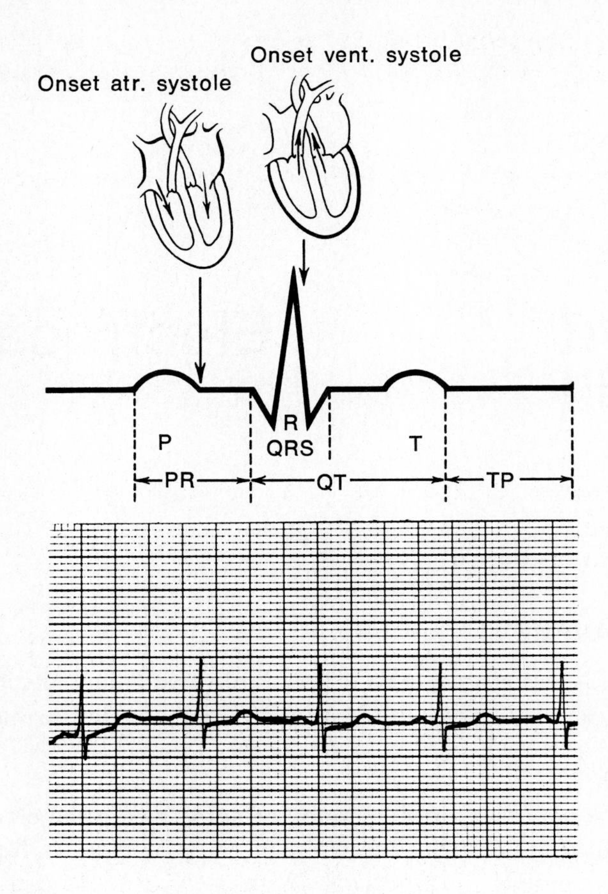

Fig. 72. Typical record of an electrocardiogram.

Electrocardiograms are obtained by means of many different types of **leads** from the body, these being deemed necessary because of the angle at which the heart lies in the chest. The three most common leads have been designated as I, II, and III. In lead I electrodes are attached to the right and left wrists, in lead II to the right wrist and left leg, and in lead III to the left wrist and left leg.

Demonstration of the principles and usage of the electrocardiograph will be given for each laboratory section. Each student will be required to have a properly labeled normal electrocardiogram in his laboratory manual.

PROCEDURE

1. Subject rolls down long stockings, socks, and so on, to the ankles. Wristwatch and/or bracelets are removed.

2. The limb electrodes are applied to the inner forearm, just above the wrist, and to the inside of the leg, just above the ankle.

3. Electrode cream, jelly, or saline paste is applied to the skin only where the electrode will make contact and then on the surface of the electrode plate before electrodes are strapped around the limb.

4. The electrode cables are connected to the electrode plates, making sure they are firmly secured. The other end is led into a cardiac preamplifier or to a self-contained electrocardiograph.

5. Having applied the electrodes and connected the subject to the instrument, recording can now begin. Before recording, check the following:

   a. Power switch recorder is on.
   b. There is sufficient paper in chart mover.
   c. Pen is centered on paper.
   d. Sensitivity of preamplifier has been set at 10 mm/mV (10 mm deflection of pen for 1 mV input signal).
   e. The subject is relaxed and lying quietly.

The record may be run for any length of time, but usually five or six ECG complexes of each lead are sufficient. A standard paper speed of 25mm/sec has been adopted by all electrocardiographers.

6. When recording is complete, disconnect the leads from the subject and remove the electrodes.

7. Clean off surplus electrode paste from subject and electrodes using gauze or paper towels.

COMPONENTS OF THE ELECTROCARDIOGRAM

1. Identify and letter P, QRS, and T waves. Calculate the heart rate. Measure the duration of P, QRS, and T. Measure the P–R interval (from the beginning of P to the beginning of QRS).

2. Compare characteristics of electrocardiograms of subject taken from three different leads: I, II, III.

3. Note that the wave pattern is different. Some leads favor certain components of the electrocardiogram as a result of the position of the electrode with respect to the direction of the spread of the electrical activity and recovery of the heart.

4. Figure 73 illustrates normal electrocardiograms taken from three different leads.

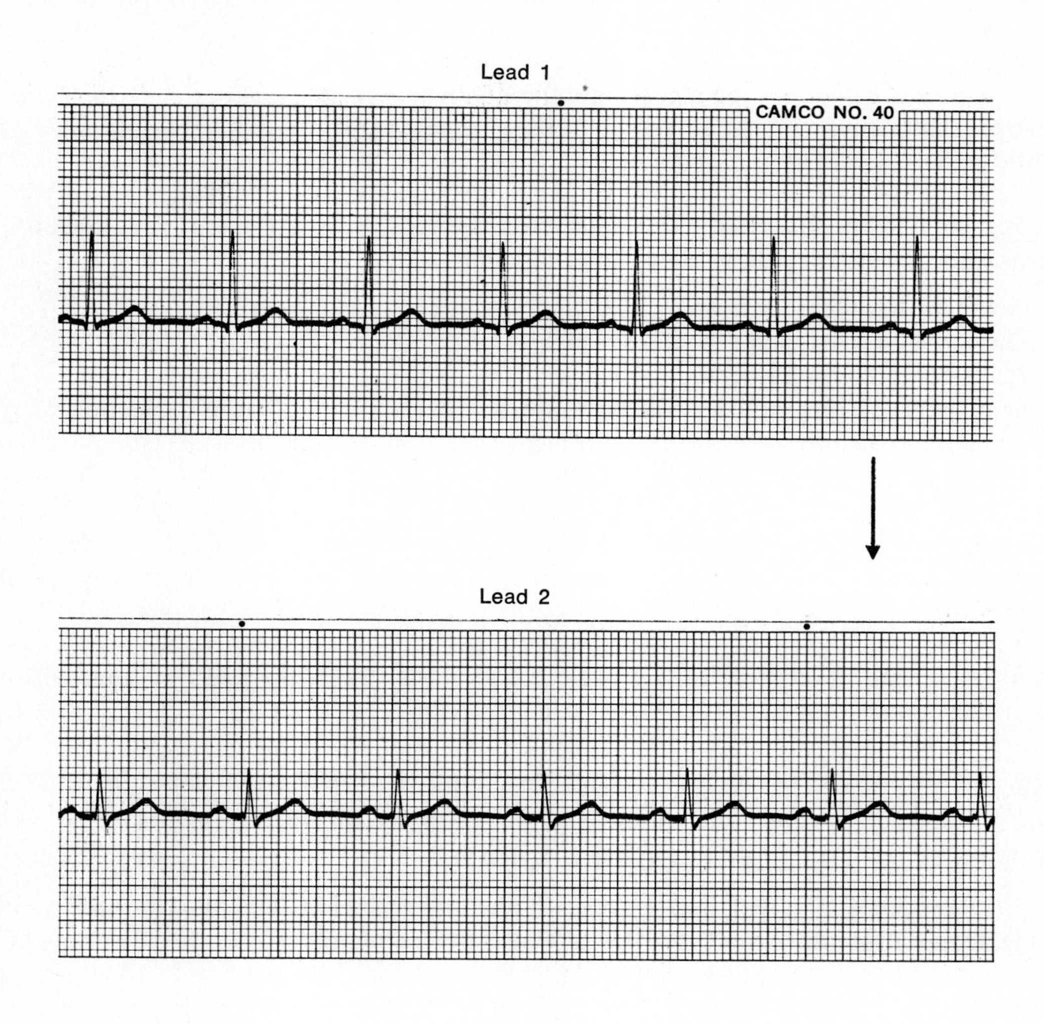

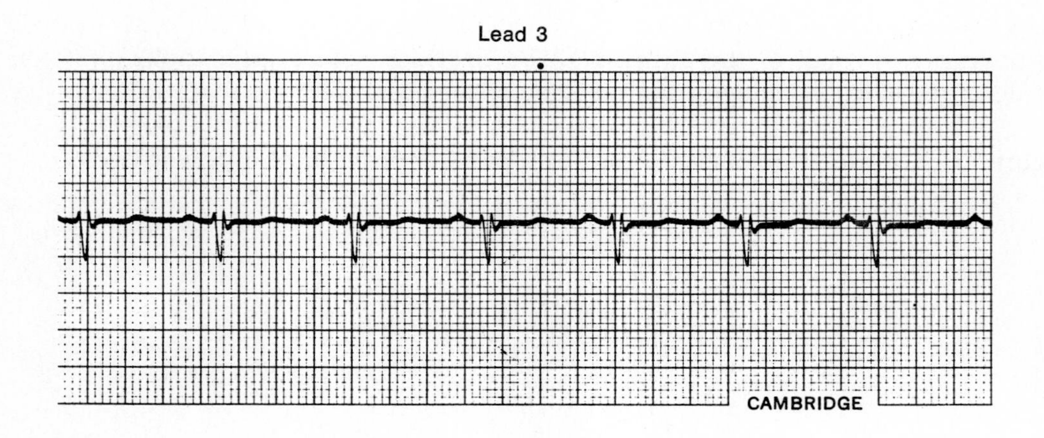

Fig. 73. Leads I to III electrocardiogram.

ELECTROCARDIOGRAPHIC REPORT

Subject____

Date____________________________ Age____________

Rate: Atria__________ Vent.__________ Subject Position__________

Intervals: P–R__________ QRS__________ R–T__________

Lead I

(Paste in strip of electrocardiogram)

Lead II

(Paste in strip of electrocardiogram)

Lead III

(Paste in strip of electrocardiogram)

Experiment XXIV | Heart Sounds

The two sounds associated with the beating of the heart are the systolic and the diastolic sounds, which may be phonetically described as "lupp" and "dupp." The cause of the systolic sound (lupp) may be explained as follows: during the ventricular systole, the ventricular valves are suddenly thrown into rapid vibration; these vibrations are communicated to the blood, heart muscle, and thoracic wall and can be heard by placing a stethoscope against the thorax. Moreover, it has been found that a skeletal muscle during its contraction produces a sound or noise. If a stethoscope is placed on a contracting muscle of the arm, a sound is heard. When the ventricular musculature contracts, a similar rumbling noise is produced, which fuses with the sound caused by the vibrations of the closing atrioventricular valves.

The second or diastolic sound (dupp) is caused by the vibration of the closing semilunar valves. When the strength of the heart beat is increased, the first sound is accentuated; an increase in the aortic blood pressure has this effect on the second sound.

If the atrioventricular valves are prevented from closing, the higher pitched sound of the systolic sound disappears, but the lower rumbling part persists. The pause between the systolic and the diastolic sounds is shorter than that following the diastolic sound.

The apex beat is that beat of the heart which can be felt by placing a finger in the fourth or fifth intercostal space, about 1 or 2 in. from the sternum. The two major causes may be (1) the hardening of the heart muscle and (2) a change in the shape of the heart from a flattened structure during diastole to a rounded one.

APPARATUS

Stethoscope, quiet room.

PROCEDURE

The apex sound just after the dupp sound and before the lupp sound is heard after diastole, and it is thus that the systolic and diastolic sounds can be placed. The first sound may be heard by placing the stethoscope over the fifth left intercostal space. The second sound may be heard by placing the stethoscope over the second left intercostal space. A knowledge of the heart sounds is essential to understand the condition and functional abilities of the pulmonary and the semilunar valves.

Experiment XXV

Circulation of Blood Through the Veins

For the blood to return to the heart after passing through the capillary bed, it must overcome the force of gravity. This would be quite a difficult job if it were not for the fact that the body provides various means by which the venous blood is aided in its return to the heart. List some of the factors that facilitate venous circulation.

PROCEDURE

Make the following observations upon yourself and your partner.

1. Notice the veins in the arm as they hang down at your side. Are they prominent and easily seen? The back of the hand will show this much better. Now raise the arm above the head. What change has taken place? Explain.

2. Determine the venous pressure in the following way. Hold your hand at the level of the heart. Have your partner raise your hand slowly until the veins collapse (be sure that the muscles in the arm are relaxed). Measure the height to which the arm is raised in millimeters, which will give you the pressure expressed in millimeters of water. Convert this into millimeters of mercury and compare with arterial pressure.

3. **Valve function in veins.** Apply a fairly tight bandage around arm of subject above elbow. Notice the veins of the forearm and hand. Are there swellings at various intervals? Place a finger at a point on a vein farthest away and with another finger press the blood forward beyond the next swelling toward the heart. Does the vein fill up again? Now remove the finger and observe what happens. Try this again, but press blood in the opposite direction. What happens? Discuss your observations.

Experiment XXVI | Circulatory Responses in Blood Vessels of the Mesentery of the Rat

Effect of Nerve Stimulation and Drugs on Circulation

All blood vessels except capillaries contain smooth muscle and receive motor nerve fibers from the autonomic nervous system. Circulatory adjustments are effected by neural and chemical mechanisms that change the caliber of the arterioles (the terms **vasoconstriction** and **vasodilatation** are generally used to refer to changes in caliber of blood vessels) and vary the rate and stroke output of the heart. **Parasympathetic stimulation** (vagus nerve) results in decreased heart rate, constriction of bronchioles, dilation of the arterioles in the skin and abdomen and small intestine, constriction of blood vessels in the heart and skeletal muscles and increased digestive activity. **Sympathetic stimulation** results in increased heart rate, dilation of the bronchioles and blood vessels supplying the heart and skeletal muscles, constriction of blood vessels supplying the skin and viscera and decreased digestive activity. Norepinephrine (adrenalin) is the chemical released at the sympathetic nerve endings that brings about the responses described. (For a complete discussion of adrenergic and cholinergic functions see Grollman's text – *The Human Body, Its Structure and Physiology,* pages 153, 156.)

In resting tissue, most of the capillaries are collapsed. However, in active tissue, the arterioles and precapillary sphincters dilate, allowing a greater blood flow into the capillaries and thereby raising the intracapillary pressure to a value above its critical closing pressure. Relaxation of the smooth muscle of the arterioles and precapillary sphincters is caused by action of dilator substances formed in active tissue (rise in carbon dioxide tension, rise in temperature, lactic acid, histamine, kinins), and possibly also by a decrease in the activity of vasoconstrictor nerves by an activation of the depressor area of the vasomotor center. Epinephrine and norepinephrine are constrictor substances.

MATERIALS AND EQUIPMENT

Ether jar, rats, scalpel, scissors, inductorium, microscope, adrenalin solution (0.005 mg/ml).

PROCEDURE

Anesthetize the animal with ether as in experiment XIV (page 157).

1. Expose the right vagus nerve and right jugular vein in the following manner:
   a. Lay the anesthetized animal ventral side up. Lift the skin in the neck region and make a midventral cut from the lower lip to the sternum. Carefully loosen the skin laterally and mop up all capillary bleeding at once with cotton pledget. Clear away any fat and connective tissue that may be masking the external jugular veins. The jugulars are located deep to the side and lie on top of the sternocleidomastoid group of muscles midventrally.
   b. Separate the band of sternohyoid muscle to expose the trachea and larynx, which lie immediately underneath these muscles (Fig. 74). Locate the common carotid artery, which lies deep in the region running alongside the trachea. The vagus nerve is usually associated with the common carotid and runs parallel with it. Place a ligature around the nerve and tie off. This will facilitate its location for later use. Place a pad moistened with warm saline solution over the exposed tissue to prevent drying and then proceed to the exposure of the mesenteries.

2. Lift the skin with fingers or forceps over the abdomen and make a midline cut exposing the abdominal viscera. Gently pull a loop of intestine out of the incision, and with the rat elevated to the height of the stage of a microscope (use books covered with plastic sheeting), place the mesentery so that it will lie across the light opening. Under low power find a field in the mesentery that gives a good view of the circulation. Notice the difference in blood flow through the arterioles, capillaries, and venules. Keep the mesentery moist with warm saline.

3. Inject 0.005 mg/kg (1000 g of body weight) of adrenalin into the jugular vein. Observe the action of this drug on the arteriole and its effect on blood flow through the capillaries. How long did the effect last?

4. Cut the previously isolated vagus above the ligature and stimulate the peripheral end of the vagus with platinum electrodes, using the inductorium set for moderate strength multiple stimuli. Observe the effect of vagal stimulation on blood flow. Vagal fibers are cholinergic. The effect is usually observed at the instant of application of the stimulus, that is, a speeding up of blood flow because of vasodilation. Continued stimulation will cause the blood flow to stop because of the fall in the general blood pressure and inhibition of heart action. Record and discuss your results.

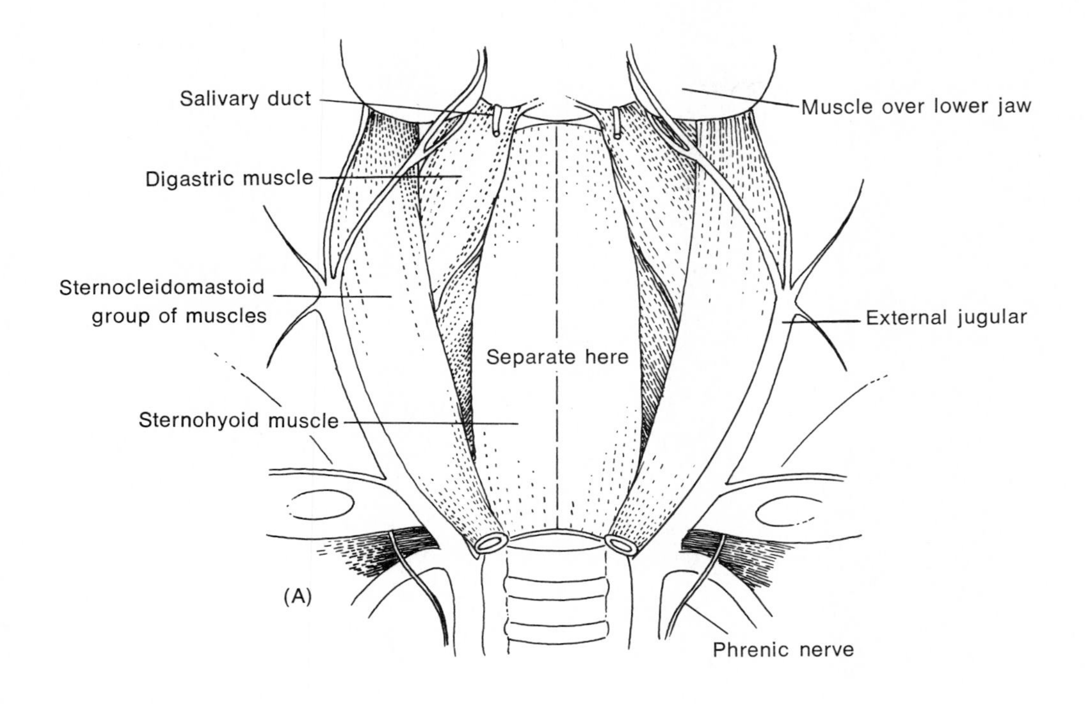

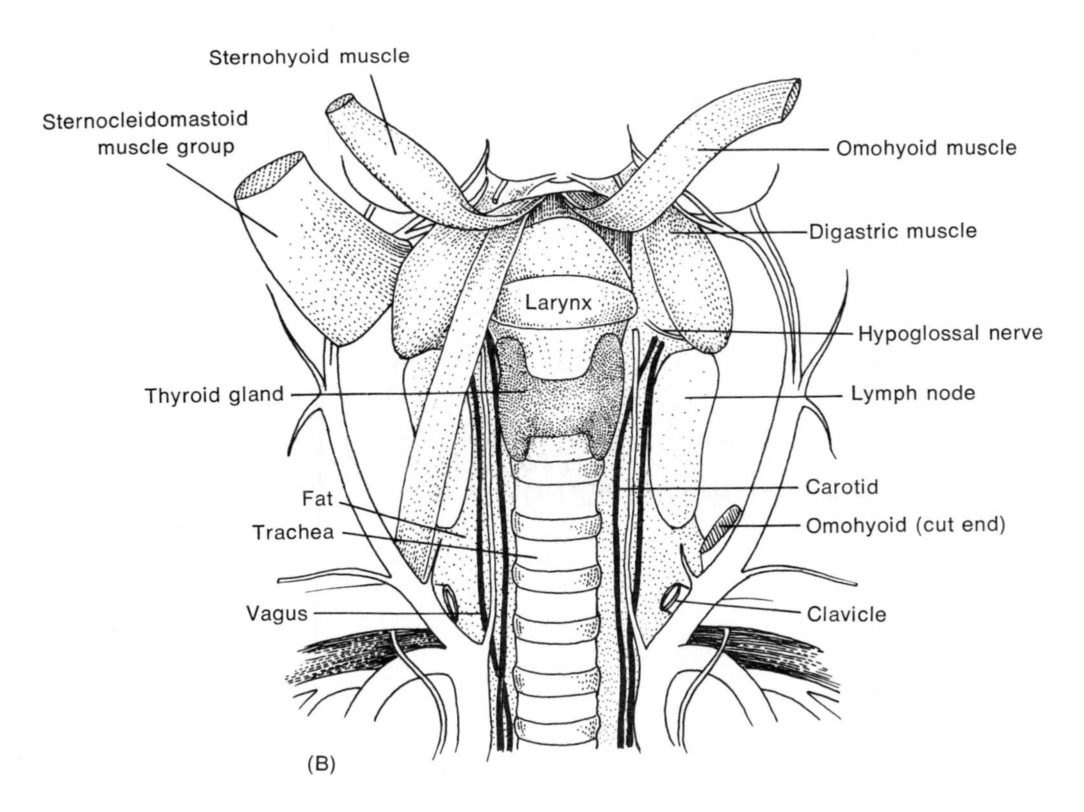

Fig. 74. Dissection guide for the exposure of the vagus nerve and jugular vein of the rat.

Experiment XXVII | Chest Measurements During Respiration

The thorax forms a type of extensible box. Sheets of muscles, the internal and external intercostal, move the ribs up as we inspire and down again as we expire. Simultaneously, the great dome shaped muscle of the diaphragm, forming a movable floor for the thoracic cavity, contracts and sinks with each inspiration, and relaxes and rises again on expiration. The synchronized movements of ribs and diaphragm enlarge the thorax considerably in all three dimensions during inspiration, and reduce it again during expiration. The lungs fill and empty with the same rhythm. It is possible to demonstrate the change in size by measuring the circumference of the chest at inspiration and at expiration.

Make the following observations by using a tape line and calipers. (tailor's tape is available at all dime stores. Wood calipers may be purchased from Keuffel and Esser, Adams and Third Sts., Hoboken, New Jersey, 07030.)

1. Measure the circumference of the chest in centimeters at two levels:

    a. At the axillae, that is, pass the tape line around the chest as high up as possible under the armpits

    b. Also, at the level of the xiphoid process (end of sternum)

2. With calipers measure the anteroposterior diameter of the chest at the nipple line; also with lateral diameter.

RECORD DATA UNDER FOUR CONDITIONS

1. With chest wall at rest at the end of quiet inspiration
2. At the end of quiet expiration
3. At end of forcible inspiration
4. At the end of forcible expiration

	Quiet Breathing		Forced Breathing	
	Inspiration	*Expiration*	*Inspiration*	*Expiration*
Circumference at axillae				
at xiphoid				
Anteroposterior diameter				
Lateral diameter				

Experiment XXVIII | Respiratory Capacity

The quantity of air drawn into and expelled from the lungs in quiet respiration is only a fraction of that which can be inhaled or exhaled during deep breathing. The air that passes in and out during ordinary respiration is spoken of as the **tidal air**. The average man after he has completed an ordinary expiration can inhale, by making the deepest inspiration possible, a large quantity of air. This air is termed the **complemental air** or **inspiratory reserve**. If starting again at the end of an ordinary respiration, a forcible expiratory effort is made, a considerable quantity of air can be farther expelled, and this is termed the **supplemental air** or **expiratory reserve**. The volumes of the complemental and supplemental airs, that is, the total amount of air that can be expelled after a maximal inspiration, are called the **vital capacity**.

These air volumes may be determined by the use of a spirometer. Sterilze the mouthpiece with alcohol. Practice exhaling through the mouthpiece with the nose shut and notice movement of indicator. Indicator reads in liters. One liter is equal to 1000 cc. Determine the various air volumes of each member of your group; record data on at least ten individuals.

Subject	*Sex*	*Height*	*Tidal Air (cc)*	*Complemental Air (cc)*	*Supplemental Air (cc)*	*Vital Capacity (cc)*
1						
2						
3						
4						
5						
6						
7						
8						
9						
10						

A fairly close relationship between height and vital capacity has been demonstrated; so much so that, by multiplying the height in centimeters by 25 for men, 20 for women, and 29 for athletes, the vital capacity in cubic centimeters can be estimated. Compare your values determined with the spirometer with those determined by estimation.

Subject	*Height*	*Factor 25, 20, 29*	*Estimated Vital Capacity*	*Vital Capacity from Spirometer*	*Differ*
1					
2					
3					
4					
5					
6					
7					
8					
9					
10					

Count the rate of quiet respiration in a fellow student who does not know that he is being observed. Find out his tidal air and calculate the amount of air respired during quiet respiration in 1 min, 1 hr, 1 day. How much of the expired air is carbon dioxide? Calculate the amount in liters and in grams.

Expired air contains 4 per cent carbon dioxide.

One liter of carbon dioxide weighs 1.968 g.

How much oxygen in liters and in grams is consumed?

Inspired air contains 20.95 per cent oxygen.

One liter of oxygen weighs 1.428 g.

RECORD DATA ON FIVE SUBJECTS

Subject	*Respiration Rate*			*Amount of Oxygen Consumed*						*Amount of CO_2 Liberated*					
	1 Min	*1 Hr*	*1 Day*	*1 Min*		*1 Hr*		*1 Day*		*1 Min*		*1 Hr*		*1 Day*	
				l	*g*	*l*	*g*	*l*	*g*	*l*	*g*	*l*	*g*	*l*	*g*
1															
2															
3															
4															
5															

Sample Calculations

Rate of respiration is 25 times per min.

Tidal air volume is 500 cc.

1 liter = 1000 cc.

Calculations:

1. **Volume of air respired per min**

$$25 \text{ breaths/min} \times 500 \text{ cc/breath} = 12{,}500 \text{ cc or } 12.5 \text{ liters/min}$$

2. **Volume respired per hour**

$$25 \times 500 \times 60 = 750{,}000 \text{ cc or } 750 \text{ liters/hr}$$

or

$$12.5 \text{ liters} \times 60 \text{ min/hr} = 750 \text{ liters}$$

3. **Volume respired per day**

$$25 \times 500 \times 60 \times 24 = 18{,}000 \text{ cc or } 18{,}000 \text{ liters/day}$$

or

$$750 \text{ liters} \times 24 \text{ hr/day} = 18{,}000 \text{ liters/day}$$

4. **Volume of O_2 consumed per day**

21 per cent of air breathed in is O_2.

16 per cent of air breathed out is O_2.

Thus, 21 - 16 equals 5 per cent that must have been consumed.

$$18{,}000 \text{ liters} \times 0.05 = 900 \text{ liters } O_2 \text{ consumed}$$

5. **Volume of CO_2 expired per day**

0.04 per cent of air breathed in is CO_2.

4 per cent of air breathed out is CO_2.

Thus, 4 - 0.04 = 3.96 per cent that must have been expired.

$$18{,}000 \text{ liters} \times 0.0396 = 712.8 \text{ liters } CO_2 \text{ expired}$$

6. **Weight of O_2 consumed per day**

Since 1 l of O_2 weights 1.428 g, then 900 liters of O_2 per day would weight 900 × 1.428 = 1285.2 g.

Or set up a proportion:

$$1 \text{ liter} : 1.428 \text{ g} :: 900 \text{ liters} : x \text{ g}$$

$$x = 900 \times 1.428 = 1\,285.200$$

7. **Weight of CO_2 expired per day**

$$1 \text{ liter } CO_2 : 1.968 \text{ g} :: 712.8 \text{ liter} : x \text{ g}$$

$$x = 1402.8 \text{ g}$$

Experiment XXIX | Carbon Dioxide Tension of Alveolar Air

CARBON DIOXIDE TENSION OF ALVEOLAR AIR

When the gases of the inspired air come in contact with the alveolar membrane, they pass through the membrane into the blood or from the blood into the alveoli by physical diffusion in accordance with the difference in pressure of that particular gas on either side of the membrane. The gas pressure in the blood as well as in the alveoli is usually expressed in terms of partial pressure of that gas, that is, pO_2 or pCO_2. In human beings the carbon dioxide tension in the alveolar air varies, but is approximately 40 mm Hg at end of expiration just prior to inspiration. A fall of alveolar carbon dioxide tension is of considerable diagnostic significance, because in cases of acidosis the tension may fall to 8 to 10 mm Hg as the body attempts to rid itself of excessive acid.

Alveolar air samples can be collected for analysis by having a subject expire forcibly through a short length of rubber hose. At the end of a forced expiration a sample of air is extracted from the tubing with a hypodermic syringe. The air sample is then analyzed for its content of carbon dioxide by a method devised by W. M. Marriott.[1]

The method of analysis depends on the fact that, if a current of air containing carbon dioxide is passed through a solution of sodium carbonate or bicarbonate until the solution is saturated, the final solution will contain sodium bicarbonate and dissolved carbon dioxide. The reaction of such a solution will depend on the relative amounts of the alkaline bicarbonate and the acid carbon dioxide present. This, in turn, will depend on the tension of carbon dioxide in the air with which the mixture has been saturated and will be independent of the volume of air blown through, provided saturation has once been obtained. High tensions of carbon dioxide change the reaction of the solution toward the acid side. Low tensions have the reverse effect; hence the reaction of such a solution is a measure of the tension of carbon dioxide in the air with which it has been saturated. The reaction of such a solution (acid–alkaline) may be determined by adding to it an indicator such as phenolsulfonphthalein, which shows definite color changes with different concentrations of hydrogen ions. Thus a certain color indicates a certain reaction.

[1] W. M. Marriott, "The determination of alveolar carbon dioxide tension by a simple method". *Journal American Medical Association,* **66**, 1594 (1916).

Solutions of a given reaction (hydrogen ion concentration) may be prepared by mixing acid and alkaline phosphates in definite proportions. Such solutions may be kept unaltered for long periods of time and can be used as standards for comparison (see page 256).

APPARATUS

La Motte Alveolar Air Carbon Dioxide Kit.* (See pages 255-256 for preparation of standards and solutions if the kit is not available.)

PROCEDURE

1. **Collection of sample for analysis.** The glass mouthpiece is placed in the tube of the rubber bag, and about 600 cc of air is forced into the bag by means of the atomizer bulb. The rubber tube is then clamped off by the pinchcock. The subject should be at rest and breathing naturally. At the end of a normal expiration, the subject takes the tube in his mouth; the pinchcock is released and the subject's nose closed by the observer or with a nose clamp. The subject breathes back and forth from the bag four times in 20 sec, emptying the bag at each inspiration. Breathing more frequently will not greatly alter the results. At the end of 20 sec, the tube is clamped off and the air analyzed. The analysis should be carried out within 3 min, because carbon dioxide rapidly escapes through rubber.

2. **Analysis of sample.** Into one of the test tubes place about 2 or 3 ml of the indicator mixture. Now insert the capillary tube into the tube of the rubber bag and force air through the solution until it is saturated, as shown by the fact that no further color change occurs. The tube is then stoppered and the color immediately compared with that of the standard color tubes.

3. **Making the reading.** Place the tube containing the test sample in the middle hole of the comparator block. In the holes on either side of the test sample place two consecutive standards (for example, 30 and 35) that seem most nearly to match the color of the test sample. View the tubes through the comparator slots, holding the etched glass toward a window or other source of daylight, and change the color standards, if necessary, until the color of the test sample exactly matches one of the standards or lies between the colors of two consecutive standards. If an exact match is obtained, the value is read off directly from the standard with which the match is obtained. If, however, the color of the test sample lies between the colors of two consecutive standards, the value is taken as the average of the two. For example, if it lies between 30 and 35, the reading is taken as 32.5.

4. **Calculation.** The standard color tubes are marked to indicate the carbon dioxide tension in mm Hg, and readings can be estimated to about 2 mm.

   a. Record readings of all groups in class.
   b. How do carbon dioxide tension and alkali reserve relate to one another?
   c. Using your text as reference, construct a table comparing the gaseous content of alveolar air at the end of inspiration and at the end of expiration. What does this reveal?
   d. How is carbon dioxide normally carried by the blood?

*La Motte Chemical Products Co., Baltimore, Md. (not available as of 1974).

5. **Results.** In normal adults at rest, the carbon dioxide tension in the alveolar air determined as described above, varies from 40 to 45 mm. Tensions between 30 and 35 mm are indicative of a mild degree of metabolic acidosis. When the tension is as low as 20 mm, the individual may be considered in imminent danger. In coma, associated with metabolic acidosis, the tension may be as low as 8 or 10 mm. Conditions other than acidosis may affect the carbon dioxide tension. Excessive pulmonary ventilation from whatever cause will lower the alveolar carbon dioxide tension; however, in these circumstances a respiratory alkalosis is produced. Thus the measurement of pCO_2 provides a useful but not complete assessment of the respiratory portion of acid/base balance. Coupled with the measurement of blood pH, one is able to evaluate more precisely the state the subject is in. In most instances metabolic acidosis is a common cause of low alveolar carbon dioxide tension.

Condition	*Alveolar CO_2 Pressure mm Hg.*	*Blood pH*
Normal	40–45	7.35–7.45
Metabolic acidosis (acid excess or base deficit)	15–35	6.8–7.35 (low)
Respiratory alkalosis (carbonic acid deficit)	10–35	7.45–7.70 (high)

Depression of the respiratory center from whatever cause will lead to an increase in the carbon dioxide tension. A respiratory acidosis (carbonic acid excess). The CO_2 tension is high (45-100 pCO_2).

Alveolar air collected as described above is essentially air which has come into equilibrium with the venous blood in the pulmonary capillaries. The tension of carbon dioxide is approximately that in the venous blood, thus, we can relate carbon dioxide pressure to volume per cent of carbon dioxide in the blood (cm^3 CO_2 in 100 ml of blood). The chart on page 256 relates carbon dioxide pressure with volume per cent of carbon dioxide in venous blood.

Preparation of Alveolar Air Carbon Dioxide Kit (Marriott's Method)

COLLECTION OF SAMPLE FOR ANALYSIS

A rubber bag of approximately 1500 cc capacity (a basketball bladder or a hot water bag can be used – if a hot water bag is used the neck can be closed off with a rubber stopper carrying a short glass tube 3/8 in. in internal diameter) is connected by means of a short rubber tube to a piece of glass tubing with fire polished ends, 1½ in. long and 3/8 in. in internal diameter. About 600 cc of atmospheric air is blown into the bag with an atomizer bulb, and the rubber tube is clamped off with a pinchcock. For collection of air sample from subject, follow same procedure as described under collection of sample for analysis using La Motte Alveolar Air Carbon Dioxide Kit.

PREPARATION OF STANDARD SOLUTIONS

Phosphate Solutions for Standard Color Tubes

1. Acid potassium phosphate solution: 9.078 g KH_2PO_4 are dissolved in distilled water, 200 ml of 0.01 per cent phenolsulfonphthalein solution is added, and the whole is made up to 1 liter.

2. Alkaline sodium phosphate solution: 11.876 g $Na_2HPO_4 \cdot 2H_2O$ are dissolved in water, 200 ml of 0.01 per cent phenolpthalein solution are added and the whole made up to 1 liter.

3. The two solutions are mixed in the proportions given in the table that follows. Different concentrations of hydrogen ion are produced that can be related to CO_2 pressure.

Acid Phosphate Solution (KH_2PO_4), ml	*Alkaline Phosphate Solution ($NA_2HPO_4 \cdot 2H_2O$), ml*	*Pressure mm Hg, pCO_2*	*Vol. Per Cent CO_2*
17.8	82.2	10	15.0
25.2	74.8	15	22.5
31.0	69.0	20	30.0
35.7	64.3	25	38.0
40.5	59.5	30	45.0
45.0	55.0	35	53.0
47.0	53.0	40	60.5
50.2	49.8	45	68.0

4. The solutions thus made are put in small test tube (10X 75 mm) and stoppered or sealed off. The standard tubes should be stored in a dark place when not in use. Each tube should be permanently marked with the pressure of carbon dioxide represented by the proportions of solutions used – which can be obtained from the chart.

Standard Bicarbonate Solution. The standard bicarbonate solution used in analyzing a sample of alveolar air is prepared by weighing out 0.530 g of sodium carbonate or by measuring 100 ml of 0.1 *N* sodium hydroxide into a 1 liter volumetric flask. Add 200 ml of 0.01 per cent phendsulfonphthalein and the whole is made up to the 1 liter mark with distilled water.

ANALYSIS OF SAMPLE

Same as given for the La Motte alveolar kit. The standard solutions described are so prepared as to give correct results when the determination is carried out at a temperature of from 20° to 25°C (from 68° to 75°F). When the room temperature is considerably higher or lower it is advisable to immerse the standard tubes in water at approximately 25°C during preparation of sample tube. The tubes are removed from the water for the color comparison. The comparison with the standard tubes can be facilitated by having a box similar to that used with the Sahli hemoglobinometer, but containing three holes instead of two. By the use of this device, slight color changes can be detected, and the temperature of the tubes is not raised by the heat of the hands. In addition, the whole device, with tubes, may be easily immersed in the water bath if it is required.

MAKING THE READING AND CALCULATION OF RESULTS

Same as described in the procedure for the La Motte Air Carbon Dioxide Kit.

RECORD RESULTS OF YOUR GROUP:

Experiment XXX | Respiratory Center

PROCEDURE

1. The subject breathes deeply for a short period of time in order to overventilate the lungs. Does the subject find that it is increasingly difficult to breathe as time goes on? What other subjective symptoms are present? Explain. If forced breathing is continued for 5 min or more, the subject will find that he has no desire to breathe. This condition is referred to as apnea. Explain.

2. The subject breathes into a paper bag, repeating the above procedure. Is it easier to breathe more deeply than before? Can deep breathing be carried on longer in this case? Does apnea set in? Explain.

3. After a quiet respiration determine how long you can hold your breath. Determine how long you can hold your breath after thoroughly ventilating the lungs. Any person not able to hold his breath for more than 20 sec is not fit for general anesthesia.

4. Determine how long the breath can be held after running in place for 2 min. Repeat after resting for about 5 min.

5. Have the subject take a deep breath and at the end of the inspiration hold. Is there an urge for inspiration or expiration?

6. Repeat, but this time hold breath after an exhalation.

7. Discuss the respiratory center, its control of breathing, and factors that may influence it.

Experiment XXXI

Human Respiratory Movements

SPECIAL EQUIPMENT

1. Pneumograph with recording tambour
2. Kymograph with special wide paper
3. Simple spirometer containing 100 per cent oxygen
4. Signal magnet operated at 1 sec intervals
5. Mercury manometer

PROCEDURE A

1. Place the pneumograph around the chest between the axilla and the xiphoid or in such a position that maximum deflections of the tambour are obtained. This will register mixed costal and diaphragmatic breathing. The subject should not be able to see the recording. (Each student in turn will act as the subject.) Connect the signal magnet for 1 sec timing and place it below the record of respirations. Note that inspiration produces a downward deflection of the writing point because the pneumograph has been lengthened, thus acquiring a greater volume, and hence lower pressure.

2. **Eupnea (quiet respiration).** Adjust drum speed so that each complete cycle occupies about 1 in. on the record. Record eupnea with the subject sitting, standing, and prone. During the sitting experiment, observe the effect of talking to the subject, ask him to read aloud, cough, or swallow. Have him make deep voluntary inspirations and expirations.

3. **Hyperpnea of exercise.** Using a slow drum speed, record standing eupnea, and while continuing the record, start running in place for 1 or 2 min. Record the effect and the recovery phase. Have the subject lie supine and record the eupnea. While continuing the record, flex and extend the subject's legs and arms passively. Record the effect and recovery.

PROCEDURE B. PERIODIC BREATHING (EXPERIMENTAL CHEYNE-STOKES); NO CARDIACS AS SUBJECTS

Record with pneumograph and tambour and 1 sec timing. *Subject should be seated and watched during the experiment, in case he loses his balance or faints.* After a few seconds of quiet respiration, breathe as deeply and as rapidly as possible for 3 min or until the symptoms become acute. Record the final 30 sec of this period and continue to record. At this point cease all respiratory effort, permitting respiration to govern its own action. (In a faint, respiration is necessarily self-governed.) Observe pulse rate throughout the experiment. Note the apnea (suspension of breathing), followed by dyspnea (labored breathing), by apnea, and again by dyspnea. Blood pressure, pulse rate, and consciousness wax and wane with the respiration. Four or five such cycles should appear before breathing begins to smooth out to normal.

PROCEDURE C. REBREATHING (CARBON DIOXIDE DYSPNEA)

Using the pneumograph, record for 10 sec intervals, every 30 sec. During recovery phase, record continuously. The spirometer, arranged for nonabsorption of carbon dioxide, is filled with oxygen. With a noseclip in place, hyperventilate, as in Procedure B above, into the spirometer for 3 min, or until the discomfort is unbearable. Note that in spite of hyperventilation, no apnea occurs because carbon dioxide is not being blown off but is actually accumulating and causes the dyspnea. Now breathe into the spirometer (with oxygen) without making a voluntary effort to hyperventilate. Note the involuntary increase in breathing. Compare these results with those in Procedure B in the same subject.

PROCEDURE D. BREATH-HOLDING TESTS

1. Breathe quietly (recording with slow drum), then hold the breath as long as possible. Record the effect and the recovery.

2. Exhale maximally, then take a deep breath and hold as long as possible. Record throughout. "Breaking point" is 30 to 120 sec (average 68 sec). At breaking point, alveolar air contains 7 to 10 per cent carbon dioxide. Excess carbon dioxide causes the break by stimulation (or raising the excitability) of the respiratory center.

3. Breathe deeply and rapidly for 30 sec and repeat test 2. Breath can now be held longer because carbon dioxide has been driven to a lower level in the blood by the hyperventilation.

4. Take a deep breath of oxygen from the spirometer before doing test 2. This should delay breaking point less than maneuver test 3. Why? Maximal breaking points are achieved by hyperventilating in pure oxygen before test 2.

5. **"Endurance," or "40 mm," test (Flack test).** Place a clip on your nose. Record pneumographically. Exhale deeply and inhale maximally. Apply your lips to the rubber tube attached to the Hg manometer and raise the Hg column to 40 mm. Maintain as long as possible, *supporting cheeks with both hands so that they take no part in the process.* The average holding time is 52 sec. While holding, observe the subject's pulse rate until breaking point is reached. Unfit persons show an earlier breaking point and a rise in pulse rate and blood pressure at an early stage. Fit subjects show no appreciable increase in pulse rate until the breaking point. Record the recovery phase. This test estimates the extent of abdominal pooling of blood, which should be small. When it is large, the intraabdominal pressure forces out blood and increases venous return, with resulting increased heart rate and arterial blood pressure.

PROCEDURE E. MAXIMUM FORCE OF EXPIRATION

Following a deep inspiration, exhale against the Hg column. A height of at least 110 mm should be obtained.

ANALYSIS

1. In Procedure A3, what would you have to do to prove that the effect is not caused by a "humoral" factor?

2. Consider the factors operating in Procedure B to cause (1) apnea, (2) dyspnea, (3) smooth breathing. Compare this with the situation in Cheyne–Stokes breathing observed in stagnant anoxia.

Experiment XXXII | Effects of Respiration on Blood Pressure in Man

SPECIAL EQUIPMENT

1. Sphygmomanometer and tambour
2. Cuff and rubber bulb
3. Mercury manometer
4. Pneumograph and tambour

PROCEDURE A

1. **Experimental arrangement.** Respiration, blood pressure from the sphygmomanometer, and the time in seconds are recorded. The cuff is applied to the upper arm and the cuff pressure adjusted to give *maximum amplitude* of oscillation of the writing point (about 15 mm above the diastolic pressure). Variations in the arterial pressure will cause changes in the pulse amplitude under the cuff. A drop or rise in blood pressure will register as a drop or rise in the mean level (base line) or the pulse wave, and usually as a decrease in its amplitude.

2. **Eupnea.** Using this cuff pressure, make a record of normal, quiet, costal respirations. The blood pressure waves produced by this breathing are simple respiratory waves. Note any change in the heart rate. The general tendency is a decrease in blood pressure during inspiration and an increase in blood pressure during expiration. The decrease in blood pressure during inspiration is attributed to the fact that

   a. The lowered intrathoracic pressure reduces the direct pressure on the intrathoracic arteries and

   b. It also increases the volume of the pulmonary bed, causing a delay in the return of the blood to the left heart. The increase in blood pressure during expiration is attributed to a reversal of these effects.

3. **Deep breathing.** With moderately deep breathing, these effects are exaggerated and additional ones may be observed. Breathe moderately deeply at approximately the same rate as that of quiet breathing. It is found that the blood pressure decreases during the early part

of inspiration (as with quiet breathing), but later (3 to 4 sec) in the inspiratory phase the pressure rises. This delayed rise in blood pressure is attributed to

a. Increased venous return to the right heart, owing to decreased intrathoracic pressure;

b. The increased blood volume finally arriving (via pulmonary circuit) brings Starling's law into play and the stroke volume is increased;

c. The Bainbridge reflex, owing to increased venous volume accelerates the heart;

d. Increased heart rate in response to decreased afferent impulses from the pressoreceptors (which responded to the initial decrease in aortic and carotid blood pressures).

During expiration these effects are reversed. *Obviously, by changing the rate of breathing, these time relations of blood pressure to respiration may be altered, or even completely reversed.*

PROCEDURE B. EFFECT OF CLOSED GLOTTIS

To exaggerate the effect of ordinary deep breathing, place the pneumograph around the chest, take a deep breath, close the glottis, and forcibly exhale, without opening the glottis. This is known as Valsalva's experiment. Positive intrathoracic pressures of 100 mm Hg may be produced in this manner. Blood pressure readings will show an increase preceded by a temporary fall in systolic pressure (normal response). Rises in the mean arterial pressures are usually less than 5 per cent above the normal levels determined in the sitting position. Overshoots of pressure above this are considered significant.

Experiment XXXIII | McKesson Metabolar and Basal Metabolism

In determining the basal metabolic rate of patients, the Benedict–Roth respiratory apparatus or McKesson metabolar is usually employed. These are closed circuit methods in which oxygen enriched air is breathed and the consumed oxygen is measured. These methods are quite simple technically and, when only total metabolism must be determined, are equally satisfactory. In the closed circuit method an average respiratory quotient of 0.82, corresponding to a caloric value for oxygen of 4.825 kcal/liter is assumed.

The apparatus consists essentially of an enclosed bellows, or bell, which holds a volume of oxygen of from 4 to 5 liters at full capacity, and a small well that contains a carbon dioxide absorbent. Inlet and outlet valves are connected to the oxygen supply.

The patient breathes through this tube. The air first passes through the carbon dioxide absorbent so that as oxygen is consumed the bellows will close upon itself as the volume of oxygen decreases, or the bell will lower. The apparatus is provided with a kymograph that is self-timed, marking in minute periods. A pen attached to the counterweight traces the respiratory excursions and shows the decrease in volume of the bellows or the spirometer bell, if the Benedict–Roth apparatus is used, as oxygen is consumed.

Figure 75 illustrates the method of drawing the oxygen consumption line and of measuring the rise of this line for any 6 min period that may be selected on the tracing.

The total heat production or energy expenditure of the body is the sum of that required merely to maintain life, together with such additional energy as may be expended for any additional activities. This minimum is called **basal metabolism**. It represents the lowest level of energy consonant with life. A measurement of the rate of heat production under basal conditions is called the **basal metabolic rate**, or **BMR**.

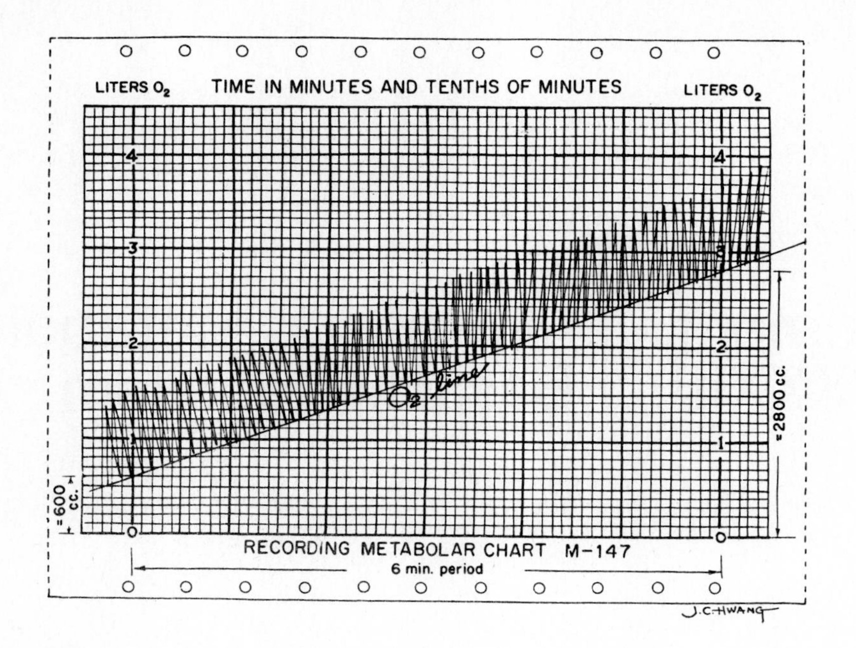

Fig. 75. Graphic record of a metabolism test: male, age 35, height 170 cm, weight 70 kg, surface area 2.0 m^2. Barometric pressure 745 mm Hg. Average temperature 21°C. Rise in oxygen line in 6 min 2800 – 600 = 2200 cm^3 or 2.2 liters in 6 min. In 1 hr, 22 liters. Correction for STP (standard temperature and pressure). See sample calculation.

CONDITIONS NECESSARY FOR MEASUREMENT OF THE BMR

1. A postabsorptive state: patient should have had nothing by mouth for 12 hr.
2. Mental and physical relaxation; usually ½ hr of bed rest is used.
3. Recumbent position during the test.
4. Patient awake.
5. Environmental temperature of between 20 and 25°C.

FORMULA FOR CORRECTION FOR STANDARD TEMPERATURE AND PRESSURE

Sample Calculation

oxygen consumed in 1 hr = 22 liters
barometric pressure = 745 mm Hg
temperature = 21°C

$$V_{STP} = 22 \times \frac{P}{760} \times \frac{273}{273 + t}$$

$$= 22 \times \frac{745}{760} \times \frac{273}{294}$$

$$= 22 \times 0.980 \times 0.928$$

$$= 22 \times 0.909$$

$$V_{STP} = 19.9 \text{ liters } O_2/\text{hr}$$

MEASUREMENT OF BASAL METABOLISM

In clinical practice, the BMR can be estimated with accuracy by measuring the oxygen consumption for two 6-min periods under basal conditions. This is corrected to standard conditions of temperature and pressure (laboratory instructor will explain). The average oxygen consumption is multiplied by 10 to convert it to an hourly basis and then multiplied by 4.825 kcal, the heat production represented by each liter of oxygen consumed. This gives the heat production in calories per hour. This is converted to calories per square meter body surface per hour by dividing calories per hour by patient's surface area.

The formula for calculating surface area is the **DuBois' surface area formula***

$$A = H^{0.725} \times W^{0.425} \times 71.84$$

where A = surface area in square centimeters
H = height in centimeters
W = weight in kilograms

Surface area in square centimeters divided by 10,000 equals surface area in square meters.

In practice, a nomogram, which relates height and weight to surface area, is used; it is based on the DuBois formula (Fig. 76).

CALCULATION OF BMR

The normal BMR for an individual of the age and sex of the patient is obtained from standard tables. The actual rate is then compared to the average normal and the rate expressed as a plus or minus percentage of the normal

Example of BMR below normal value. A male, 35 years of age, 170 cm in height, and 70 kg in weight, consumed 1.2 liters of oxygen in 6 min (this has been corrected to STP 0°C, 760 mm Hg).

$$1.2 \times 10 = 12 \text{ liters } O_2/\text{hr}$$

$$12 \times 4.825 = 58 \text{ kcal/hr}$$

$$\text{surface area} = 1.81 \text{ m}^2 \text{ (from the DuBois formula)}$$

$$\text{BMR} = 58 \text{ kcal}/1.81 = 32 \text{ kcal/m}^2 \text{ hr}$$

The normal BMR for the sex and age of this patient by reference to the DuBois standards is 39.5 kcal/m² hr. His BMR, which is below normal, is then reported as

$$\frac{39.5 - 32}{39.5} \times 100 = 18.5 \text{ per cent or minus } 18.5$$

*DuBois and DuBois, "The Measurement of the Surface Area of Man," *Archives Internal Medicine,* **15**: 868 (1915).

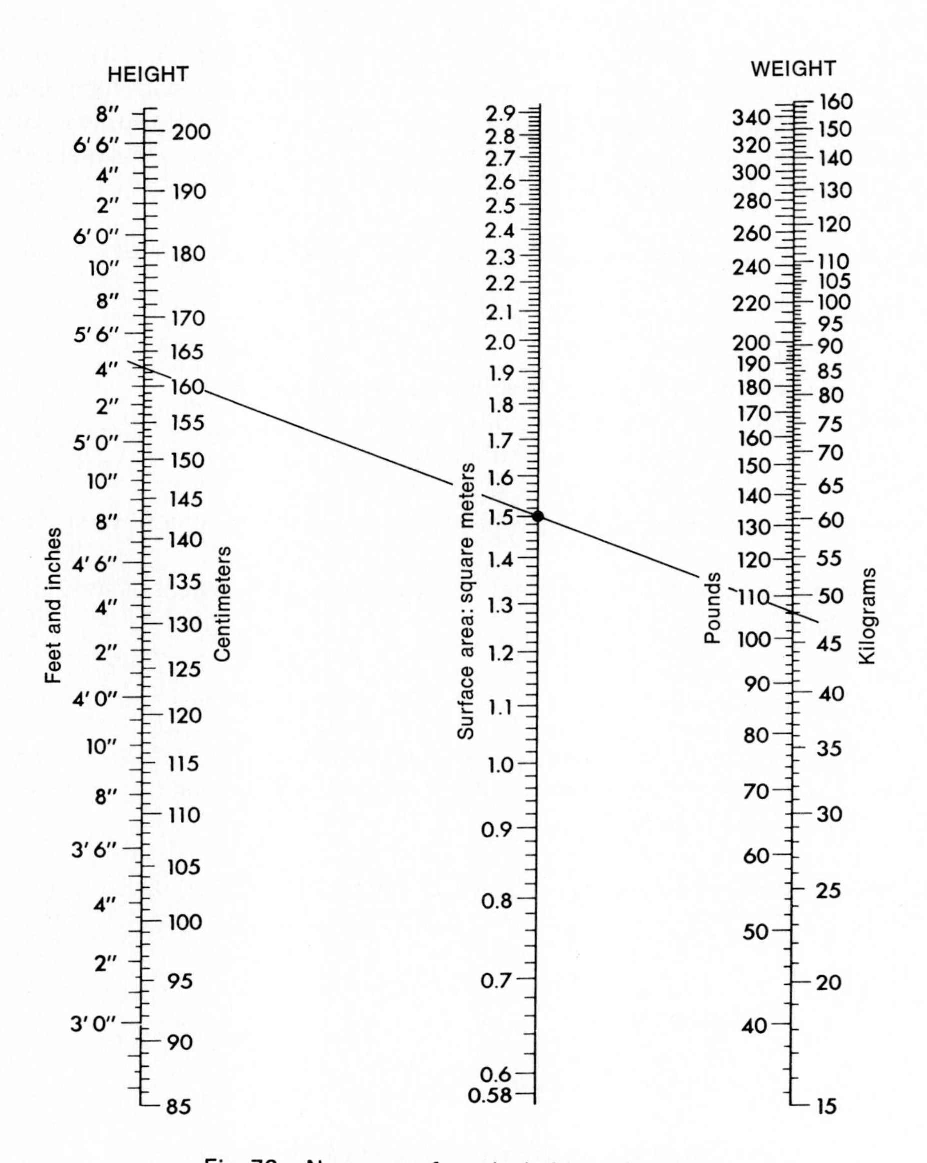

Fig. 76. Nomogram for calculating surface area.

A BMR between - 2 and +20 is considered normal. In hyperthyroidism the BMR may exceed +50 to +75 per cent. The BMR may be - 30 to - 60 per cent in hypothyroidism.

Conversion Factors

1 lb = 0.455 kg   1 in. = 2.54 cm

Use nomogram to obtain surface area.

Metabolar Respiration Record

Name____________________ Case No.__________

Address__________ Hour__________ Date__________

Age_____ Sex_____ Weight_____ Height_____ Surface area__________

Temp._____ °F__ Resp. rate__________ Rhythm__________

Mental attitude: calm, nervous, very nervous.

S________

Last food at__________ o'clock______ Blood pressure D________

PP________

Pulse rate__________ Rhythm__________

Basal metabolism________ % or__________ kcal/m^2 hr

Diagnosis____________________

Pathology____________________

Treatment and results____________________

Basal Metabolism Calculation

Barom.________ mm Mercury

O^2 absorbed____ liters in____ min

O_2 consumption per hour________ liters

________ Factor Table 1

Temp. of apparatus________ °C

Muscular Movement { none / slight / much

Results { satisfactory / questionable / unsatisfactory

Basal metabolism Rate________ %

1 per hr corrected

Surface area (Table 2)

________ kcal/hr m^2

Normal kcal output (Table 3)

Experiment XXXIV | Salivary Digestion

The greater part of the food elements in the diet requires special treatment to render it capable of absorption and utilization by the body. The necessary changes are chiefly hydrolytic in character and involve especially the action of enzymes found in the different parts of the gastrointestinal tract. The first such action occurs in the mouth when food materials are mixed with **saliva**, the secretion of the salivary glands. The principal enzyme of the saliva is **salivary amylase or ptyalin.** Ptyalin is a mixture of enzymes that catalyze the splitting of starch into simpler molecules. The first product of the action of the enzyme on starch is the formation of soluble starch which in turn is gradually split into smaller fragments known as dextrins. Soluble starch will give you a blue color with iodine; the dextrins give a red color with iodine (erythrodextrins). If splitting continues to free maltose, no color reaction will occur.

Action of Saliva on Starches

PROCEDURE

The progress of this enzymatically catalyzed reaction will be followed by observing (1) the disappearance of the reacting substrate (starch) using the iodine test and (2) the appearance of the product of the reaction (maltose) using the Benedict's test. These two tests will be run simultaneously.

Place about 10 drops of starch paste (1 per cent) into a test tube, add about 20 drops of saliva and stir thoroughly. Note the opalescence of the starch–saliva mixture. Remove a drop of the solution to a spot plate and test with iodine. At the same time remove about 5 drops of the starch–saliva solution to one of a series of 3 test tubes containing 5-ml portions of Benedict's reagent and set aside. At 1-min intervals remove a drop of the solution to a spot plate and test by the iodine test. The opalescence of the starch solution should soon disappear, indicating the formation of **soluble starch**, which gives a blue color with iodine. The soluble starch should soon be hydrolyzed into **erythrodextrins**, which give a red color with iodine. When the erythrodextrin stage has been reached, as indicated by the iodine test, remove a second sample of the solution (5 drops) to the second tube of Benedict's reagent and set aside. Erythrodextrins are then hydrolyzed into **achromodextrins**, which give no color with iodine. This is the achromic point of the reaction. When the achromic point has been reached, remove a third sample of the solution to the third test tube of Benedict's reagent. Complete the Benedicts' tests by placing the three tubes in a boiling water bath for about 5 min and noting the degree of reduction in each tube. Tabulate your results with the iodine test and Benedict's test in parallel columns.

EXPERIEMENTS ON SALIVA

A satisfactory method of obtaining saliva for experiments is to chew a small piece of pure paraffin wax or a rubber band and collect the saliva, which has been stimulated, into a beaker.

1. **Microscopic examination.** Examine a drop of saliva under the microscope after staining with methylene blue. Try to identify the following: **epithelial cells, salivary corpuscles** (small oval cells containing many fine granules), **fat droplets**, leukocytes, and bacteria.

Make a drawing of the field labeling any of the above constituents you may find in the saliva.

2. **Reaction.** Test the reaction of saliva to litmus or other suitable indicator paper. Estimate the approximate pH.

3. **Test for mucin.** To a small amount of saliva in a test tube, add 1–2 drops of dilute acetic acid. Acetic acid precipitates proteins in the saliva – mucin being the most abundant protein present.

4. **Biuret test.** Add a few drops (2–5) of a Biuret reagent solution in a test tube of water. The formation of violet color indicates a positive test for protein.

Benedict's test. After boiling the test mixture for 5 min, as outlined previously, allow the test solution to cool spontaneously (do not hasten by immersion in cold water).

In the presence of a reducing sugar the entire body of the solution will be filled with a precipitate which may be red, yellow, or green in color, depending upon the amount of sugar present. In the presence of over 0.2 to 0.3 per cent glucose, the precipitate will form quickly. If no glucose is present, the solution will remain perfectly clear.

Biuret test. Reagent: 1 per cent copper sulfate solution is added drop by drop with constant stirring to some 40 per cent sodium hydroxide solution until the mixture assumes a deep blue color.

The mixture may be used by adding it drop by drop to the test solution, with mixing, until the solution assumes a violet color. Alternatively, 2 or 3 drops of the reagent may be permitted to flow down the side of the inclined tube, in which case the reagent forms a layer beneath the solution and the violet color appears at the interface between the two liquids if the protein is present. If the protein has been digested a pink color results.

REAGENTS

Starch solution (1%). Suspend 1.0 g starch in 100 ml distilled H_2O. Gradually heat to boiling while stirring. Cool to room temperature.

Iodine solution for iodine test. Prepare a 2 per cent solution of potassium iodide and add sufficient iodine to color it a deep yellow. A drop or two of this solution in a small amount of starch solution is sufficient.

Benedict's solution. Benedict modified the Fehling solution to produce an improved reagent which has largely displaced the latter in routine laboratory practice.

copper sulfate	17.3 g
sodium citrate	173.0 g
sodium carbonate	100.0 g
distilled H_2O to make	1 liter

With the aid of heat, dissolve the sodium citrate and carbonate in about 800 ml water. Pour through a filter, if necessary into a glass graduate, and make up to 850 ml. Dissolve the copper sulfate in about 100 ml water. Pour the carbonate-citrate solution into a large beaker, add the copper sulfate solution slowly with constant stirring, and make up to 1 liter. The mixed solution is ready for use and does not deteriorate upon long standing.

Experiment XXXV | Gastric Digestion

After mastication, the food is carried by peristaltic movements of the esophagus to the stomach. Here it undergoes further mechanical disintegration and chemical changes primarily in the protein constituents. The food thus treated is in a better condition to be handled by the intestines to which it is passed on in small portions at a time and in which digestion is completed.

Gastric juice contains the enzymes pepsin, rennin, and lipase. The acidity of the gastric juice is caused by free hydrochloric acid. Pepsin, the most characteristic of the enzymes of the gastric juice, acts upon proteins to split them down into smaller fragments.

PROCEDURE

1. Introduce a protein material such as fibrin or coagulated egg white into a capillary tube. Place the capillary tube into a flat-bottom vial containing a few milliliters of pepsin solution (0.5 per cent) add 0.4 per cent HCL, dropwise, until pepsin solution becomes acid as indicated by a blue color with Congo red paper. Incubate the mixture at 40°C for 2–3 days. The original protein has been digested to proteoses and peptones.

2. Apply the Biuret test. A positive test is indicated by a pink color which means that proteoses and peptones are present.

3. Conditions essential for the action of pepsin. Prepare six flat-bottom vials as follows:

   1. 5 ml pepsin solution
   2. 5 ml 0.4% hydrochloric acid
   3. 5 ml pepsin hydrochloric acid solution (enough acid to keep mixture acid to Congo red paper)
   4. 2 or 3 ml pepsin solution and 2 to 3 ml 0.5% sodium carbonate solution
   5. Sodium carbonate (2 to 3 ml of 0.5%)
   6. Distilled water (2 to 3 ml)

Into each vial introduce a capillary tube containing protein and place tubes in the incubator of water bath at 40°C for ½ hr carefully noting the changes that occur.

4. Digestion of protein (fibrin) in a pepsin hydrochloric acid solution is indicated first by a swelling of the protein because of the acid, and later by a disintegration and solution of the fibrin because of the action of pepsin. If uncertain at any time whether digestion has taken place, apply the Biuret test to the filtrate. A positive reaction (pink color) signifies the presence of proteoses and peptones, the presence of which indicates that digestion has taken place.

5. Now combine the contents of tubes 1 and 2 and see if any further change occurs after standing at 40°C for 15 to 20 min. Explain the results obtained from these experiments.

Materials such as pepsin and fibrin may be purchased from Nutritional Biochemicals Corp., Cleveland, Ohio 44128.

Experiment XXXVI | Pancreatic Digestion

PROCEDURE

1. Prepare the enzyme solution by adding 5 ml of 1 per cent pancreatin solution to a test tube. Add 0.5 ml of 10 per cent Na_2CO_3 and 4.5 ml of water. Mix well and divide equally into three flat-bottom vials. Into one vial introduce a capillary tube containing coagulated egg white. Into the second vial introduce a capillary tube containing fibrin. Set aside the third vial of enzyme solution for the last part of the experiment. Incubate the two vials at 35–40°C for ½ hr. Check for protein digestion by observing any changes within the capillary tubes as well as performing the Biuret test on the solution in the vial. Did protein digestion occur? Which protein digests more readily, the egg white or the fibrin? Why?

2. Add 1 ml of starch paste to the vial of enzyme solution set aside earlier. After about 10 min, test a drop or two of the mixture for starch using the iodine test. If starch is still present, allow the digestion to continue until the achromic point has been reached. Once the achromic point has been reached, test the solution for reducing sugars using the Benedict test (add 5 drops of the solution to 5 ml of Benedict's reagent).

3. Discuss your results.

Experiment XXXVII | Urinalysis

Normal urine is a highly complex solution of many organic and inorganic compounds representing largely waste products derived from the metabolic processes. The principal dissolved substances of urine are urea, uric acid, creatinine, ammonia, and the chlorides, phosphates, and sulfates.

In pathological conditions, proteins, sugars, acetone, bile, and hemoglobin may be excreted and form elements such as casts, and red and white blood cells may be present. Thus a properly performed urinalysis can give considerable information not only regarding the condition of the urinary tract, but concerning other structures and functions of the body as well. Urine specimens will be brought to the laboratory in suitable containers by individual students.

Note color, transparency, odor, and reaction. Approximate pH by use of litmus paper.

TESTS

Specific gravity. The specific gravity is conveniently estimated by the urinometer. Each type of urinometer is adjusted to give accurate readings at a definite temperature marked on the stem. Corrections of 0.001 are added to the urinometer reading for every 3°C above the standard temperature for the urinometer employed; likewise, 0.001 is subtracted for each 3°C below the temperature.

Place the urine in a cylinder. Immerse the float in the urine without touching the sides or bottom of the cylinder. When the float comes to rest, the reading is taken, from the bottom of the meniscus.

Normal 1.003 to 1.030

Albumin. The term **albumin**, as applied to urine, refers to both serum albumin and serum globulin. Urine tested for albumin must be clear. Simple filtration through ordinary filter paper usually suffices to clear a turbid specimen. Heat 5 ml of urine in a test tube to boiling. A precipitate forming at this point is caused by albumin or phosphates. Acidify with three to five drops of acetic acid, adding the acid drop by drop to the hot urine.

If the urine remains clear, no albumin is present. A precipitate that clears upon the addition of acid is caused by phosphates. One caused by albumin persists and may become more flocculent. The test may be used to estimate roughly the quantity of albumin present by allowing the precipitate to settle overnight. Precipitates reaching to ½, ⅓, ¼, and ⅒, the height of the urine column represent roughly 1.0, 0.5, 0.25, and 0.1 per cent albumin, respectively. Carry tests out on own urine as well as on specimens provided by the laboratory.

Glucose. Normal urine contains traces of glucose and other reducing substances too small in amount to respond to the ordinary tests. At times, however, various sugars may appear in the urine and give positive reactions. Glucose is the most common and of the greatest clinical importance.

Most tests depend on the reduction of alkaline solutions of copper to cuprous oxide. Benedict's is the most satisfactory.

Albumin, when present in appreciable amounts, should be removed by acidifying the urine with dilute acetic acid, boiling, and filtering. The test is then performed on the filtrate.

REAGENT—BENEDICT'S QUALITATIVE SOLUTION

cupric sulfate crystals	17.3 g
sodium carbonate (anhydrous)	100.0 g
sodium citrate	173.0 g
distilled water to make	1000.0 ml

PROCEDURE

Place 5 ml of reagent in a test tube. Add eight drops or 0.5 ml or urine (no more). Boil vigorously for 1 to 2 min over an open flame. If a large number of tests are to be run, it is more convenient to place tubes in a boiling water bath for 5 min. Allow the tubes to cool slowly.

RESULTS

If the mixture remains clear, or a white turbidity develops (ureates or phosphates), no sugar is present. In the presence of glucose, the entire solution becomes opaque and filled with a precipitate, which may be green, yellow, orange, orange brown, or bright red, depending upon the quantity of sugar present. Results are reported as positive or negative. Positive results are recorded as

Weak positive – green turbidity indicating 0.1 to 0.25 per cent glucose
Moderate positive – yellow to orange precipitate indicating 0.5 to 1.0 per cent
Strong positive – heavy orange or red precipitate indicating over 1.5 per cent

Experiment XXXVIII | Endocrinology

Estrus Cycle

The female rat passes through a cycle of sexual activity about every 5 days. Although this cycle differs from the menstrual cycle in several ways, the same type of hormonal mechanism is in operation. Much of our present knowledge concerning the regulation of sex cycles was gained from study of the rat.

MATERIALS AND EQUIPMENT

Microscope, slides, swabs, Shorr single differential stain.

PROCEDURE

1. Each table is assigned young adult female rats (100 to 120 g). By use of daily vaginal smears, the stages of the estrus cycle can be determined.

Obtaining a vaginal smear

a. Prepare a swab by wrapping a small amount of cotton around the end of a toothpick.
b. Moisten the swab with saline and gently insert into vagina.
c. Rotate the swab slightly; withdraw and press upon a clean slide.
d. Quickly fix in ether-alcohol.* (Proceed to step b of Method of Staining Vaginal Smears, and continue with steps for staining.)

Method of Staining Vaginal Smears

a. Spread the aspirated fluid thinly upon the slide. (Do not allow to dry.)
b. Fix in ether-alcohol for 1 min.
c. Apply stain by a dropper; leave on for 1 min.
d. Dehydrate by dipping 10 times into 70 per cent and then 95 per cent alcohol.
e. Blot slide and clear in xylol.

*If the smear is not to be stained, press the swab into a drop of saline placed on the slide.

f. Mount by placing two to three drops of isobutyl methacrylate dissolved in xylol upon slide. Spread by tilting the slide and allow to dry in a level position.
g. When viewing the slide, study the thin areas. The cornified cells are brilliant orange and the noncornified are blue green.

Composition of Stain*

ethyl alcohol 50 per cent	100 ml
Biebrich scarlet (water sol.)	0.5 g
orange G	0.25 g
fast green FCF	0.075 g
aniline blue (water sol.)	0.04 g
phosphotungstic acid	0.5 g
phosphomolybdic acid	0.5 g
glacial acetic acid	1.0 ml

2. Sacrifice one of the rats during estrus and one during anestrus. Examine the uteri of these two animals and observe difference in size, color, and fluid content.

Stages of Estrus Cycle

VAGINAL CYCLE

Stage 1. Lips somewhat swollen, mucosa slightly dry, eight to twelve layers thick, stratum corneum appearing beneath surface layer, no leukocytes. *Smear:* uniform sized epithelial cells only. (Average length – 12 hr.)

Stage 2. Lips swollen, mucosa dry and lusterless, cornified layer now superficial and beginning to detach, no leukocytes. *Smear:* cornified cells only. (Average length – 12 hr.)

Stage 3. Lips sometimes still swollen, mucosa dry and lusterless, cheesy substance in lumen, cornified layer completely detached, no leukocytes. *Smear:* large numbers of cornified cells only. (Average length – 15 hr.)

Stage 4. Swelling of lips gone, mucosa slightly moist, greatly reduced in height, infiltrated with leukocytes. *Smear:* cornified cells and leukocytes. (Average length – 6 hr.)

Stage 5. Mucosa moist and glistening, thin, some leukocytes. *Smear:* leukocytes and epithelial cells. (Average length – 57 hr.)

UTERINE CYCLE

1. Uterus becomes distended with fluid toward end of this stage.

2. Reaches greatest distention in early part of stage and then regresses; vacuolar degeneration of epithelium sometimes begins.

*If the smear is not to be stained, press the swab into a drop of saline placed on the slide.

3. Epithelium undergoing vacuolar regeneration.

4. Vacuolar degeneration reaches its height and regeneration proceeds.

5. Epithelium regenerated.

OVARIAN CYCLE

1. Growth and enlargement of follicles; corpora lutea of preceding ovulation show fatty degenerative changes.

2. Large Graafian follicles; ova may undergo maturation.

3. Ovulation, secretion of fluid into periovarial space.

4. Young corpora lutea containing cavity or with cavity just closed; follicle smallest, ova in oviducts.

5. Corpora lutea normally fully formed and functional.

HEAT CYCLE

Animals mate toward end of stage 1 and most of stage 2.

Testicular Function: Spermatogenesis

The testes (Fig. 77A), like the ovaries, have a dual function of producing reproductive cells and sex hormone.

PRODUCTION OF MATURE SPERM

The production of mature sperm cells by the seminiferous tubules is referred to as **spermatogenesis** and is a process that begins at puberty and continues without interruption throughout the life of the male. Spermatozoa are formed from the primitive parent cells, called **spermatogonia,** which form the basilar layers of the seminiferous tubules (Fig. 77B). Spermatogonia undergo extensive mitotic activity to give rise to smaller spermatogonia, which are pushed closer toward the lumen of the tubule. These pass through a period of growth in which each spermatogonium undergoes considerable enlargement, becoming a **primary spermatocyte.** Each primary spermatocyte divides, forming two **secondary spermatocytes.** Previous to this division, the chromosome number characteristic of the species is present. Each secondary spermatocyte containing *half of the characteristic chromosome number* undergoes division to produce two cells called **spermatids.** Spermatids are attached to **Sertoli's cells,** which appear to act as sources of nourishment for the developing sperm cells. Each spermatid, by a process of differentiation without any division, becomes a mature sperm cell, or **spermatozoon.** Thus beginning with a primary spermatocyte, two cell divisions have occurred resulting in the formation of four functional sperm cells, each with the **haploid number** of chromosomes. The two separate mitotic divisions involved in this process constitute **reduction division,** or **meiosis** (Fig. 77C). In most mammalian species the process of mature sperm cell production requires

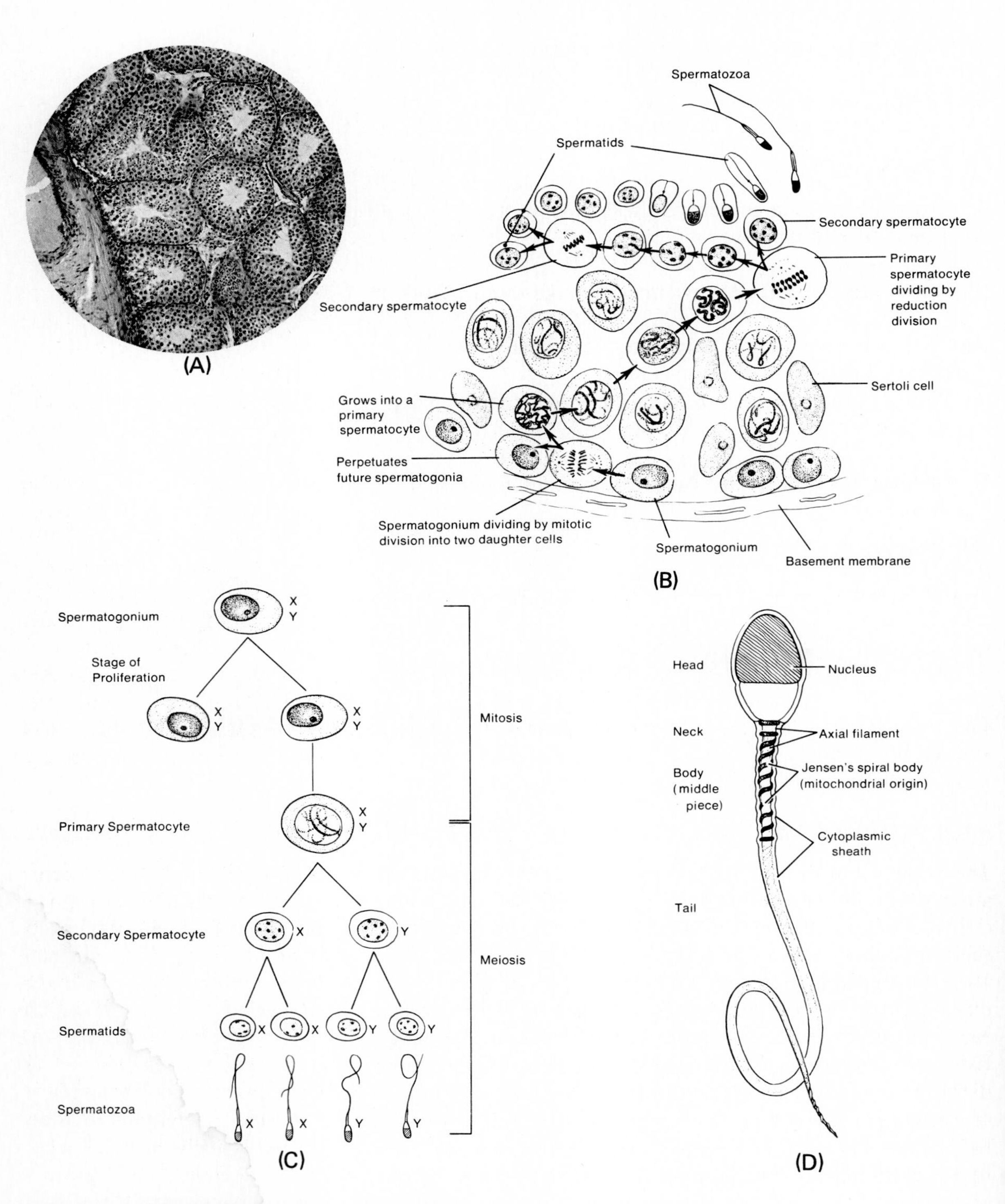

Fig. 77. Testicular function — spermatogenesis.
A. Human testis, showing seminiferous tubules. (*Courtesy General Biological Supply House, Chicago.*)
B. Schematic section of seminiferous tubule, showing sperm cells at different stages of development.
C. Spermatogenesis.
D. Mature sperm cell.

about 30 days. Since the sex chromosomes in the male are X and Y chromosomes, half the spermatozoa contain the X chromosome and half the Y chromosome.

Although there are differences in shape of mature sperm cells among different species of animals, they all share in common basic structural features such as a head, neck, middle piece, and long tail. In man the spermatozoon (Fig. 77D) measures 55 to 65 μ in length. The **head** is oval or elliptical in shape and consists mostly of nuclear material with little cytoplasm. The **neck** is a rather delicate structure, believed to be derived from the centriole of the undifferentiated sperm cell. It connects the head with the middle piece, or body, of the spermatozoon. The **middle piece** is cylindrical or spindle shaped and contains many fine fibrils grouped into a fascicle called the axial filament. A spiraled structure called Jensen's spiral body surrounds the axial filament. Enclosing the entire middle piece is a sheath of cytoplasm in which are suspended minute particles called microsomes. The **tail** is the longest structure of the mature sperm and appears to be a continuation of the axial filament of the middle piece, which tapers off, finally terminating in man in an end piece as either a single fibril or several fibrils in tassellike form. Under the ordinary light microscope only the head and a long, whiplike tail can be distinguished.

The nuclear material that makes up most of the head of the sperm cell carries the chromatin material, which consists approximately of 43 per cent DNA. At fertilization it is this material that contributes to the zygote the paternal genetic characteristics. The small amount of cytoplasm surrounding the nuclear material of the head, which is called the **acrosome**, is believed to be concerned with the release of enzymes necessary for fertilization. The middle piece is the area of metabolic activity wherein the energy released in the breakdown of glycogen is transferred to high energy phosphate compounds. The tail is the means by which the spermatozoon propels itself through its surrounding medium; that is, it is the locomotor organ of the cell. However, spermatozoa are nonmotile while in the seminiferous tubules. As sperm cells mature, they become detached from Sertoli's cells and are transported to the epididymis, where they are stored. Ciliary action of the rete tubule cells and pressure are believed responsible for the transmission of these forms to the epididymis.

The exposure of air and the addition of alkaline secretions from the prostate, seminal vesicles, and Cowper's gland that occur at ejaculation stimulate sperm motility. In the absence of the emission of spermatozoa from the epididymis, the sperm cells disintegrate and are reabsorbed by the tubules. In addition to being stored in the epididymis, sperm cells are probably also stored in the vas deferens, which ascend in the posterior part of the spermatic cord, becoming the spermatic canal, where it runs to the base of the bladder and becomes enlarged, sacculated, and narrowed, forming the ampulla, which receives the duct of the seminal vesicle. From here the ejaculatory duct conducts the sperm through the substance of the prostate gland to the penile urethra (Fig. 68A). The seminal vesicles, the prostate, and Cowper's gland produce an alkaline secretion that serves as a suspending medium for the spermatozoa. The seminal fluid is thus a complex secretion, being composed of the spermatozoa produced in the testes and the secretions produced by the structures described. It has a thick, whitish, striated appearance, and if examined microscopically, it is seen to contain innumerable motile spermatozoa.

At the time of ejaculation, the chief factor involved is the propulsive force generated by the contraction of the perineal muscles. The average volume of an ejaculation is approximately 4 ml, and in man the average sperm count is 120,000,000 sperm per milliliter. It can be seen

that in the total ejaculation there are more than a half billion sperm; since usually only one sperm fertilizes one ovum, this is another example of nature's providing an overabundance in order to protect a vital function, in this case to assure the continuation of the species. If the sperm count falls below 50 million per milliliter, good fertility is not assured. In addition, if abnormal forms of sperm cells exceed 25 per cent of the sperm population, the man will, in all probability, be sterile. Abnormal forms of spermatozoa are usually encountered in normal seminal fluid and involve only differences in size and shape of the head.
Spermatogenesis is affected adversely by both too high and too low temperatures. In human beings, as in most mammals, normal spermatogenesis cannot occur at the temperature prevailing in the body cavity but can occur at the temperature prevailing in the scrotum. Although mature spermatozoa are capable of surviving very low temperatures, the process of spermatogenesis is inhibited by exposure to cold. The scrotum, therefore, appears to be a structure serving a dual thermoregulatory function. It permits the testes to maintain a temperature 3 to 4 degrees below body temperature, which appears to be the optimal temperature for sperm production and also helps to protect the testes against too-low environmental temperature. It is the mobility exhibited by the scrotal sac that allows for this thermoregulatory control whereby an optimal temperature can be maintained for spermatogenesis.

The scrotal sac is suspended primarily by the cremasteric fascia, which is composed of a double layer of areolar and elastic tissue enclosing a definite but thin layer of striated muscle. It is principally the cremasteric fascia that affords proper retraction of the testicles and also their descent from the body in response to temperature changes. When the temperature drops or when there is danger of trauma, the muscles contract, causing the testes to be brought tight up against the body (retraction) where they can have the benefit of the body heat and protection of the body cavity. When the temperature increases, the muscle relaxes, allowing the scrotal sac to fall away from the body cavity and the body heat and thus permitting the testicles to maintain an optimal temperature for spermatogenesis.

It has been suggested that certain types of sterility in man due to spermatic arrest may be caused by the insulating effects of tight fitting, jockey type shorts and to occupations that expose the man to very high temperature for prolonged periods of time. In cases of cryptorchidism where the testes have failed to descend into the scrotum, no viable sperm are produced because of the temperature. Such men are sterile but are quite capable of sexual intercourse, and they develop all the male secondary sexual characteristics since testosterone is produced by such testes in adequate quantities.

PROCEDURE:

Examine the following microscope slides.

1. **Testis, human** – showing general structure, capsule, tubules and interstitial tissue.

2. **Testis, rat** – showing stages of spermatogenesis.

3. **Sperm smear, human** – identify all parts of mature sperm cell.

Ovarian Function: Oogenesis

The ovaries are homologous with the testes of the male, since they arise from the same tissue and from the same site in the developing embryo, and, like the testes, they serve a dual function in producing sexual reproductive cells and elaborating a complex group of sex hormones. Each ovary, which represents the female gonad, or organ in which the ova are produced, consists of a fibrous coat that encloses the stroma of the organ (Fig. 78A). The stroma consists of fibrous connective tissue containing some smooth muscle cells, especially in the area of the attachment to the broad ligament. The surface of the ovary is covered by a layer of **cuboidal cells**, which constitute the **germinal epithelium.** Within, are a number of vesicles of various sizes, each with an ovum surrounded by an epithelium. These are called **ovarian follicles.** It has been estimated that there are approximately 400,000 immature follicles in both ovaries at birth; these are called **primary follicles.** These primary follicles are produced during fetal life from the germinal epithelium and contain the cells that become ova when the female matures. These cells represent the **primitive oocyte** and are called **oogonia.**

Beginning at puberty, which is usually between the twelfth and fifteenth years of life, the primordial follicles mature, one approximately every 28 to 30 days. The oocyte within becomes a **mature ovum,** and the entire structure is known as a **graafian follicle.** Since the reproductive life of the female spans approximately 38 years (ages 12 to 50) and one ovum is developed per month during this time, it would appear that only 450 ova out of a potential 400,000 are ever released. This is another example where nature is lavish in providing a safety factor in important functions, in this case to ensure the reproductive capacity of the human species.

PRODUCTION OF MATURE OVA

The ovaries, as already mentioned, begin their major function of producing mature reproductive cells with the beginning of puberty. One mature ovum develops from a primordial follicle approximately every 28 to 30 days. Each primordial follicle consists of an oocyte, or primitive, nonmatured ovum, surrounded by a single layer of flattened epithelial cells (Fig. 78A,B). Under the influence of follicle-stimulating hormone (FSH), which is secreted by the anterior pituitary gland, the development of the mature egg (oogenesis) and a mature graafian follicle begins (Fig. 78A,C). The **oocyte** continues to increase in size, fat globules appear in the cytoplasm, and a thin transparent structure develops between the follicular cells and the oocyte called the **zona pellucida.** The enlarged oocyte, now known as the **primary oocyte,** divides unequally, forming one large **secondary oocyte** and a small cell called the **first polar body.** The size difference is not due to nuclear size difference but to the amount of cytoplasm. The secondary oocyte undergoes another division to form a large mature egg cell, or **ovum,** and a **second small polar body.** The divisions of the oocyte thus result in one functional mature germ cell and two nonfunctional polar bodies.

In immature germ cells the number of chromosomes characteristic of the species is present and is said to be diploid. Under these conditions chromosomes are represented in pairs. During the process of maturation the chromosome pairs are separated and distributed between daughter cells, with the result that in the mature sex cell only half the number of chromosomes characteristic of the species is present. The reduced number is called the **haploid number** of chromosomes. It is the halving of the chromosome number in mature sex cells that main-

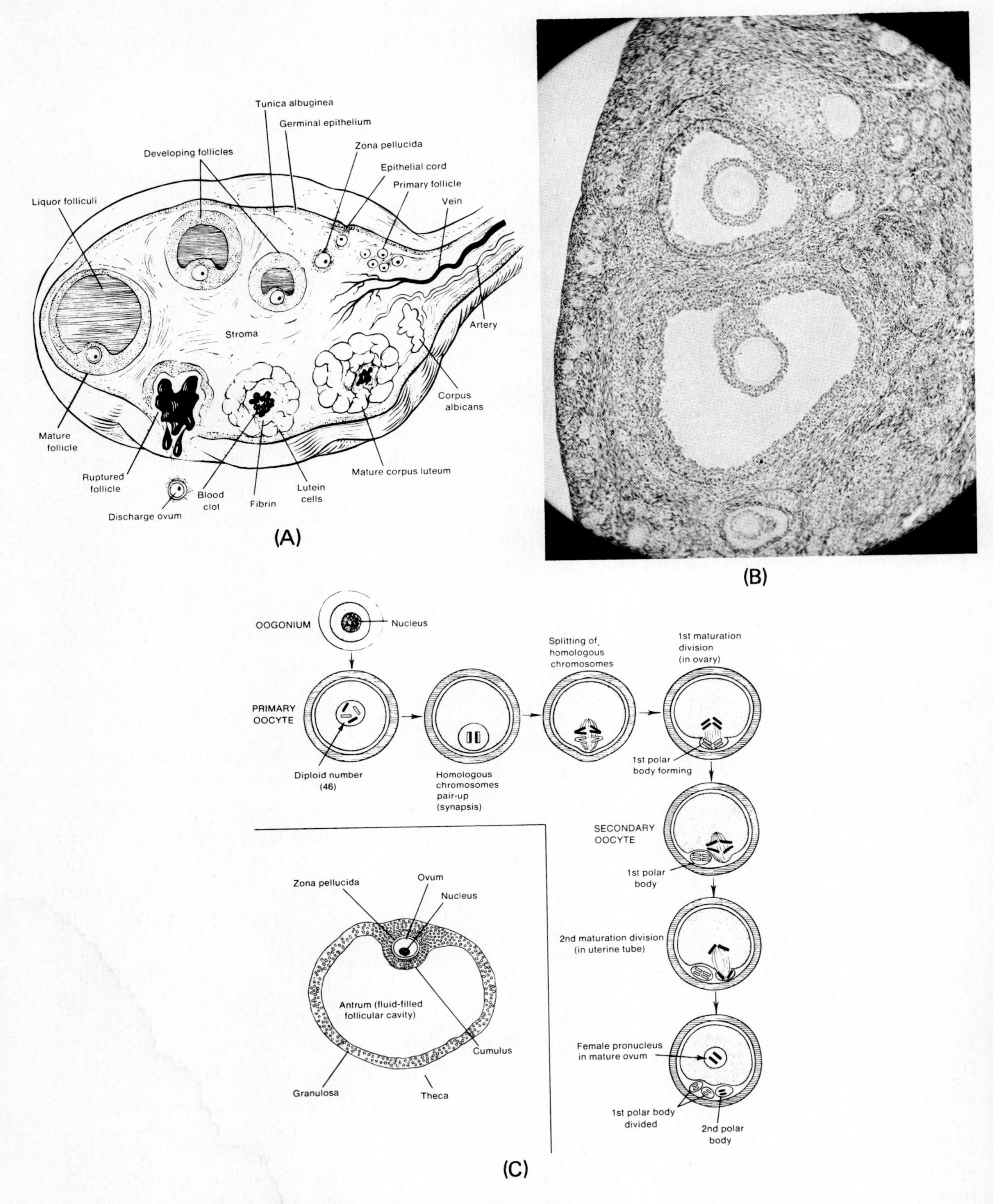

Fig. 78. Ovarian function — oogenesis.
A. Ovary and associated structures.
B. Section of ovary showing mature graafian follicle and ovum.
C. Oogenesis.

tains constant the number of chromosomes characteristic of the species. Figure 78C schematically illustrates maturation, or gametogenesis (oogenesis in the female, spermatogenesis in the male).

During meiosis, the term applied to the several divisions of the oocyte described above, the structure surrounding the egg is also undergoing change. The primitive follicular cells proliferate, forming a layer of stratified cuboidal epithelial cells called the membrana granulosa. A cavity begins to develop between the cells of the membrana granulosa, which becomes filled with a clear fluid known as **liquor folliculi.** Extending into the cavity but entirely surrounded by a mass of cells from the membrana granulosa is the ovum. This entire structure is known as the graafian follicle, which is surrounded by a connective tissue wall known as the **theca folliculi.** The theca consists of two distinct layers: the outer known as the theca externa and the inner as the theca interna.

The graafian follicle begins its development deep in the cortex of the ovary (Fig. 78A), but completes its maturation at the surface, appearing as a blisterlike bulge on the surface of the ovary just prior to ovulation. Finally, as a result of the degeneration of its wall, it ruptures, discharging the ovum, the follicular fluid, and many of the follicular cells into the peritoneal cavity. This process is called ovulation and takes place approximately midway between menstrual periods, that is, 14 ± 2 days prior to the next expected menses. Siegler* has made the observation that ovulation may occur as early as the eighth day or as late as the twentieth. Although the ovum is released into the peritoneal cavity, the fimbriated end of the fallopian tube (Fig.70B) immediately grasps it, and the ovum is propelled along, by the ciliated epithelial cells lining its lumen, to the uterus. It has been estimated that it takes the ovum approximately 72 hr to arrive at the uterus after ovulation. However, fertilization of the ovum usually occurs at the upper end of the fallopian tube and not in the eterus. If fertilization is to occur, experimental evidence indicates that it must take place within 24 hr after ovulation.

THE FORMATION OF THE CORPUS LUTEUM

Immediately after ovulation a blood clot forms at the site of the ruptured follicle; it is often referred to as the **corpus hemorrhagicum** (Fig. 78A). However, it should be mentioned that although this structure is definitely present for a number of lower animals, it is not readily found in the human female or monkey. In human beings, the ruptured site of the follicle is closed almost immediately with little bleeding. The remaining granulosal and thecal cells undergo a series of changes, becoming large and glandular in appearance and containing many yellowish granules. These cells, now called luteal cells because of the yellow granules found within their cytoplasm, replace the blood clot at the site of the evacuation of the follicle and give rise to a definite glandlike spherical organ called the **corpus luteum**, which is partly embedded in the ovarian cortex. The corpus luteum is maintained for about 14 to 15 days, although its functional life, during which it elaborates progestational hormones such as progesterone, is about 7 to 11 days. With the onset of menstruation the corpus luteum undergoes rapid degeneration because of fatty infiltration, and many of the cells are removed. It now appears as a white fibrous structure and is aptly called the **corpus albicans.** This structure atrophies and after several months sinks deep within the stroma of the ovary as a tiny scar called the **corpus fibrosum.**

*S. L. Siegler, *Fertility of Women.* Philadelphia, J. B. Lippincott Company, 1944.

If fertilization of the ovum takes place and pregnancy occurs, the functional activity of the corpus luteum may be prolonged for 5 to 6 months. The corpus luteum of the pregnant female is known as the **corpus luteum verum** in contrast to that of the luteal body of the regular cycle, which is referred to as the **corpus luteum spurium.**

PROCEDURE

Examine the following microscope slides:

1. **Ovary** – cross section of mature cat ovary showing general structure.

2. **Ovary, cat** – showing mature graafian follicle with germ hill (cumulus oophorus) (Fig. 78B).

3. **Oogenesis** – model or blackboard drawing showing the process of ovum development from a precursor cell with diploid number of chromosomes (46 for the human) to a haploid mature ovum with 23 (human) chromosomes.